H Chris Ransford

God and the Mathematics
of Infinity

What Irreducible Mathematics Says about Godhood

H Chris Ransford

GOD AND THE MATHEMATICS OF INFINITY

What Irreducible Mathematics Says about Godhood

ibidem-Verlag
Stuttgart

Bibliographic information published by the Deutsche Nationalbibliothek
Die Deutsche Nationalbibliothek lists this publication in the Deutsche Nationalbibliografie;
detailed bibliographic data are available in the Internet at http://dnb.d-nb.de.

Bibliografische Information der Deutschen Nationalbibliothek
Die Deutsche Nationalbibliothek verzeichnet diese Publikation in der Deutschen
Nationalbibliografie; detaillierte bibliografische Daten sind im Internet über http://dnb.d-nb.de
abrufbar.

Cover picture: Alexey Byckov

ISBN-13: 978-3-8382-1049-0

© *ibidem*-Verlag / *ibidem* Press
Stuttgart, Germany 2017

Table of Contents

Introduction

There is no consensus within society as to the nature of the reality we live in. Most hold that the universe was born from physical processes but don't quite agree exactly *how*, with some reckoning that a foundational Big Bang brought about by purely physical laws and events gave rise to our universe, and others suggesting alternative scenarios. Some insist that the universe was created by some ineffable Godhead but readily disagree as to *when* the act of creation took place, with common estimates and/or beliefs ranging from a mere few thousand to 14-odd billion years ago. There is no consensus either on the question of whether ours is the lone universe in existence, or if there are in fact other universes, embedded in a wider multiverse or metaverse.

Put simply, different people hold utterly irreconcilable views of what simple reality is. This lack of consensus also extends to the more abstract areas of human purpose, cause and destiny. Some subscribe to a more spiritual view and see their lives as part of a meaningful grander scheme of things, while others see life as the meaningless outcome of a long series of ultimately random events, circumscribed by laws of physics which got their start when the universe began through some foundational event, itself ultimately explainable by the laws of physics. Because this latter view of ultimately pointless lives playing themselves out in a random and purposeless universe is often perceived as repugnant, and also because many people have experienced at some point or another in their lives what they felt were deeply spiritual experiences, attempts at larger-than-life, transcendental explanations have been routinely sought throughout history. At different times and places in humankind's early history, these attempts at explanation have been retold and collated, and eventually often codified, thus giving rise to the many competing and often mutually incompatible cults and religions which have been with us since the dawn of mankind. Various religions and cults (1) still very much thrive in our modern world, and often dominate international events and narratives.

This book is an exploration of why, if there is such a thing as a Godhead, any possible objective approach to understanding It must begin by not bypassing the most factual and objective tool of analysis at our disposal bar none—provided we have first firmly established that tool's suitability and validity within the domains where we will be using it. That tool is pure mathematics.

This book first examines how and why the use of some numbers is a legitimate and indeed indispensable tool to any objective approach of both what Godhead may possibly be, and cannot possibly be, and how the mathematics of numbers can be safely used within its established areas of suitability. At first sight it would be of course natural to doubt that theology and hard science, let alone simple mathematics, would have any relevance to each other. But as I showed in recent a peer-reviewed article entitled 'Immanence vs. Transcendence: A Mathematical View', they do happen to have surprisingly direct relevance: simple math has the capability to incontrovertibly solve in a few strokes conundrums that have vexed Theologians for centuries (as the question in this paper did to wit, if there is a Godhead, is It present everywhere, or does It mostly keep to some hallowed place in space and time, some privileged Eden mostly removed and away from the rest of the universe?) Even whenever science based on simple math demonstrably *cannot* solve some theological question, as is sometimes the case, then this in itself is of huge interest. As we shall see, a remarkable instance of this latter case is the existential question itself: math by itself cannot possibly prove whether an infinite Godhead exists or not, so that the claims we sometimes hear that science either proves or disproves the existence of a Godhead cannot be supported (or more precisely, to be able to answer the existential question math would have to define the Godhead, in a definition that may not meet consensus.) The narrative briefly examines whether the attribute of *some* infiniteness should be inherent to Godhood, and then goes on to look at intriguing and often counter-intuitive results that immediately arise from using neutral mathematics.

Part 2 is a brief exploration of what math may say about variously held beliefs and assumptions. By way of illustration, it delves deeper into a few selected dogmas and tenets, said by a number of separate creeds to have been dictated directly by the Godhead. It examines in the light of mathematical analysis whether such are logically tenable or even compatible with an infinite Godhead, and whether, should we look at their logical consequences, some may not unwittingly lead to hidden contradictions and logical impossibilities. It then examines whether godlike Infinity can exist at all in any actual reality, and if it does, what role it can possibly have, and not have, in the mundane affairs of man.

Part 3 starts from an assumption that a Godhead consistent with mathematics exists, and analyses the inescapable consequences of that assumption—some of which may turn out to be quite unexpected, which will shed new light on a few old questions, including the vexed question of how, if an all-powerful Godhead exists, then why does evil still exists (math provides a straightforward and astonishing answer to that question.) It also looks in a new light at the continued prevalence of perceived mystical experiences throughout the ages.

Extensive use of end notes is made, which are used whenever some point calls for further context, buttressing or underpinning, but should its argument be kept in the main text, it would lead to a lengthy, somewhat narrowly specialized discussion and thereby risk losing the thread of the main narrative into a stray off-tangent.

A few passages in the main text are indented. They either consist of short passages quoted from outside sources, or alternatively of some brief background relevant to a point currently under discussion, but which may be however safely skipped without impairing the ability to follow the argument.

PART 1

Which Nature of Reality?

Theologians and, more recently, scientists have traditionally taken on the role of answering the question of what it all means. Their day job is to probe the ultimate nature of reality—to understand what it *is* that makes the world tick. These two communities (2) approach the issue from vastly different angles and with totally different tools, yet they share a common purpose of understanding and describing reality.

Theologians

The community of theologians goes back thousands of years, and still strongly endures. Many claim to hold special knowledge of Godhood, imparted to them through a variety of ways—meditation, divine revelation past or present, ancient scriptures, and the like. But different theologians routinely offer starkly different and, despite areas of overlap, often mutually contradictory visions of who or what the Godhead may be. Since irreconcilable views of divinity have historically led to severe social disharmony, to crimes and wars, both civil and foreign, and still do so today, some *objective* means of telling what may possibly be true or at least harbor a measure of truth from what is likely patently wrong is long overdue. Much adding to the confusion, academics with impeccable credentials, from a wide range of reputedly objective disciplines—Richard M. Gale, Michael Martin, Richard Swinburne, Victor Stenger, Peter Russell, and many others, have approached the subject from a variety of supposedly rigorously impartial angles over the years, and yet have still reached opposite conclusions with seemingly metronomic regularity—which further underscores the need for an absolutely objective tool of analysis. Could it be that, much like the faint flapping of butterfly wings may bring about inordinately big effects on distant, virtually unrelated related events—a phenomenon known as the 'butterfly effect'—the slightest subconscious bias may be stealthily determining the eventual outcome of analyses not incontrovertibly fully rooted in pure calculation-driven objectivity?

So whom, and what, are we to believe? And why and how different and incompatible views of Godhood can arise in the first place? Historically, the use of some mathematics and/or logic has been sometimes attempted: so-called ontological arguments were made by some theologians to demonstrate the existence of a Godhead. Such arguments, however, seem flawed (3). A far more compelling case for the possible existence of a Godhead has come from a far unlikelier source—a mathematician who was not in the business of seeking answers to queries of a spiritual or theological nature, but who appeared to stumble onto one: Georg Cantor, the pioneer of the formalized study of infinities, demonstrated that the mere existence of infinities ultimately leads to a stark mathematical contradiction, a full-blown breakdown of mathematics and of logic itself. He could only resolve the contradiction— which he called an *antinomy*—by positing the existence of a super-infinity, something much too vast to ever be approachable through the mathematics of infinity alone, but which required the deployment of much more than mathematics to be even remotely fathomed—an infinity approachable by us, however dimly, only if we use both our left (logical) and right (intuitive) brain. He made an argument that this super-infinity is the Godhead itself. Again, contrarily to the ontologists' approach, Cantor did not set out to find a mathematical definition or proof of Godhead, but, as he saw it, had to invoke the Godhead in order to resolve the intractable contradiction he discovered in the mathematics of infinity.

We will look more in depth at Cantor's arguments below. Generally speaking, everything about a Godhead is about infinities and infinite attributes (4), although, surprisingly, a few theologians disagree. The theologian Harold Kushner, for instance, argues in his book 'When Bad Things Happen to Good People' that the Godhead is *not* infinite. He is led to this puzzling conclusion by his analysis of the question of why a Divinity would see fit to allow 'good people' to undergo 'bad things', from his standpoint as someone who believes in a specific Godhead with precisely defined qualities and attributes, set forth in the narrowly defined framework of a rigidly established dogma. Building on lines of thought first put forward by Gersonides in the fourteenth century and more recently by others, such as Levi Olan, Kushner rather extraordinarily concludes that Godhead is in fact powerless to stop 'bad things' from happening. Under his view, the Godhead is neither infinite nor almighty. We shall keep here with the majority view that any Godhead must be infinite, and that infinity is the very quality that ultimately gives rise to the disruptive, extraordinary phenomenon of divinity. A more in-depth analysis of the question is presented in note (5).

The issue of why and how it is legitimate to use simple numbers-based mathematics in this context is a valid question, dealt with under (6). The bottom line is that some math is at the very least valid within certain areas and domains relevant to the questions at hand, and we shall restrict ourselves to such domains. By demonstrating incontrovertible facts, math will enable us to tell apart what can possibly be, and what most definitely can *not* be. It will enable us to come closer to an understanding of who or what a Godhead could possibly be, *if* there is such a thing as a Godhead. It will also help in circumscribing the existential question itself—can there possibly *be* a Godhead to begin with?

A few guidelines on how mathematics can and should be used is appropriate here, so please bear with me as I briefly set them forth here. Broadly speaking, math can only deal with precisely defined words describing sharply delineated concepts. For instance, should we say that Godhead is love, or compassion, these somewhat fuzzy concepts cannot be readily probed or analyzed by numbers or by math. But should we say that Godhead is *infinite* love, infinite compassion, then the 'infinite' part of the statement is directly amenable to mathematical analysis: indeed, math is the *only* tool available in the box that objectively deals, or even *can* deal, with infinities.

Within the framework of a number of possible limitations and provisos, math will thus allow for deploying non-subjective, logical, 'left-brain' approaches, all the more indispensable because the wonted subjective, right-brain approaches seem to unfailingly lead to contradictions. Somewhat unexpectedly, math will also turn out to be helpful in analyzing emotional and right-brain approaches, and it will demonstrate *why* contradictions inevitably arise when exclusively right-brain approaches are used.

Scientists

Scientists, especially physicists, constitute the second community whose job it is to probe and understand reality, and therefrom to explain it to the rest of us. That there happens to be no more consensus as to what it all means amongst physicists than there is amongst theologians will surprise no one, and only underscores anew the acute need for using the most objective tool of analysis bar none.

A key ongoing debate among physicists today is between the so-called Aristotelian view of reality, and the so-called Platonic view. At its core, the Aristotelian view is a materialistic view (7), whereas the Platonic view holds that the ultimate nature of reality is not materialistic, but that abstract concepts, such as, first and foremost, an underlying abstract mathematical structure, play a or indeed *the* determinant role in weaving the fabric of our reality. In different shadings, this latter view has become more popular of late, in part thanks to published works by the likes of Max Tegmark, Lawrence Krauss, and others. Of course, if a person believes in any deity, that person then *necessarily* takes the Platonic view of reality, because he or she believes in a Universe which, at the very least, simultaneously accommodates both material reality itself *plus* some spiritual, immaterial transcendent being. Indeed, as we shall see, not only modern common sense but also straightforward mathematics proves that a Godhead, if It exists, *cannot* possibly be material.

This modern view may seem self-evident today, but not so long ago the image of Godhead as some avuncular man in the sky was not uncommon. In 1961, the Soviet cosmonaut Yuri Gagarin became the first human to travel in space. The then Soviet leader Nikita Khrushchev was quoted afterwards as saying, in all seriousness, in a speech in support of the USSR's secularist policies, "Gagarin flew into space, but didn't see any God there" (a quote that was later falsely attributed to Gagarin himself.) As recently as 1971, even John Lennon saw fit to write the lyrics: 'Imagine there's no heaven, above us only sky', in apparent reference to the then still surprisingly commonly held view of a material, three-dimensional Godhead resident somewhere in space.

Using Mathematics

Whenever we fail to use a purely mathematical approach, whether from a believing or disbelieving or open standpoint, our views of the Godhead are bound to be almost equally naive. For instance, we may broadly agree on a definition of the Godhead as being infinite and disincarnate. But there are many quite different infinities, so which one is it exactly that we are favoring? We will probably naturally opt for some apex infinity—but, unless we do the math, we won't achieve anywhere near a full understanding or appreciation of the inescapable consequences that must flow from such a definition. Whenever or wherever infinities are involved, an astonishing degree of complexity kicks in and the plots thicken immeasurably, most often well beyond and differently from what we would naturally expect (8).

A math-based approach will also prove especially helpful in objectively analyzing received *dogmas*, i.e. the established foundational doctrines of many organized belief systems. Long-established dogmas are mostly accepted *as are* today, and reputed to spring from revealed truths. Any request for further justification is often staved off by a demand for a 'leap of faith', or some other demand for unquestioning obedience or acquiescence, often on the basis of the say-so of some (relatively) ancient texts deemed to be unassailable bearers of truth. As we shall however shortly see, dogmas are on the whole fully amenable to mathematical analysis.

As mentioned, math will also prove instrumental in analyzing the *consequences* of infinity—where the presence of infinity within certain contexts ineluctably leads. We cannot on the one hand accept or assert that a Godhead is infinite, and then blithely attribute traits, thoughts, or properties to that Godhead that would mathematically belie such infinity. We can hardly claim that we believe or for that matter disbelieve in something *unless* we fully know what it is that we believe or disbelieve in, including the flow-on consequences of such beliefs or disbeliefs. Whenever a contradiction that would belie infinity is uncovered, if we are to retain Its infiniteness there will be no choice but to abandon the corresponding purported trait of the Godhead.

Math will also be used to endow even everyday terms with precise meanings, and we shall endeavor to make such word definitions as broadly consensual as possible. Many concepts and/or realities tend to be loosely defined, and despite the widespread use of the same words, consensual meanings remain elusive, and different people may associate very different meanings to a same term or phrase. Illustrating the point, the age-old question 'Do you believe in God?' is utterly meaningless unless both words 'believe' and 'God' are very precisely defined, yet loose variations of this very question have historically led to all manner of strife, lethal and otherwise. We will therefore adopt here below a first definition for 'Godhead', which may then become further refined as logical analysis proceeds. Likewise, should we for instance say that a Godhead is infinitely *intelligent*, we must then find some way of defining *intelligence* appropriately accurately, so that its meaning both meets with consensus and becomes amenable to objective mathematical analysis. Most often, it will be easy to do so: for instance, the definition of 'infinitely intelligent' here would be, simply put, someone with an infinite IQ or IQ equivalent (irrespective of whatever particular IQ measure would be used, see *note* 9). Whenever some quality or attribute seems fuzzy, we will endeavour to find a way of defining it, even if provisionally so, bearing in mind that we could later on be led to a further refining of definition.

Last but not least, numbers will provide a handy way to *illustrate* certain concepts which would otherwise prove harder to apprehend— first and foremost, *infinity* itself. For instance, we can quite easily conceptualize how the unbounded series of whole numbers 1,2,3, 4,5,6...... goes on forever and never ends, and as such is infinite. Trying to conceptualize infinity, and infinities, by any other means or within contexts different from mere numbers may sometimes prove less straightforward, so that picturing infinities through the simple expedient of numbers shall provide a helpful shortcut towards visualizing infinity in a variety of other contexts.

What constitutes mathematical proof?

Two separate kinds of proof will be used in this book: first, the purely mathematical variety—proof that requires no further input from any other science to be able to stand on its own—be it physics, chemistry, psychiatry, psychology, or any other scientific discipline. A basic example of such a proof would be the statement that, "if A minus 1 is equal to 0, then necessarily, A is equal to 1." That's it—no further ado, no further discussion nor proof is needed to truthfully and incontrovertibly state that in this above case the value of A is unequivocally equal to 1, period. We'll call such proofs A-type proofs.

Examples of such A-type proofs include, for instance, the above-cited fact that if one believes in a Godhead, then one has no choice but to believe in the existence of some more complex universe or *multiverse* beyond our simple material universe—i.e., an outside reality that goes beyond the currently known material universe. Should our particular universe be finite (a question which we will revisit below), then the simple statement that a Godhead exists and has some infinite traits proves the existence of something else beyond that universe—because there is simply no mathematical scope or room to accommodate infinity within a finite universe. There are of course ways, as we shall see, whereby our very own universe could harbor infinity, so that a whole new separate universe or a metaverse or multiverse is not needed to accommodate infinity, but at the very least the existence of some infinity within the universe is in some way required if we are to accommodate divinity.

But the plot thickens, because there are many different infinities. Therefore, should we say and accept that a Godhead is infinite, or, say, *infinitely* good, what does this statement exactly mean? Are we content with some lower-ranking infinity, or shall we insist on a higher ranking, or, if such exists, on *a* or *the* apex infinity? We will examine these issues at some length in these pages.

Time is also involved: mathematically, we will see that if a Godhead exists, It by definition is the master of time and space. The only way to do so is to exist out of time, so that both the past and, crucially, the future do not hold sway on a Godhead the way they do on us, and therefore at least some parts of the wider multiverse must be timeless. But how can any place anywhere *be* timeless? We shall see below that this eventuality is much easier to conceptualize than would first appear.

Beside purely mathematical proofs, there also exists another kind of valid proof. This kind, which we'll call B-type proofs, makes essential use of math but also synergistically draws upon other scientific disciplines to work in full.

Here's an example.

One of your neighbors is a veterinarian doctor who maintains a colony of bonobo apes in his vast back garden. He is authorized to do so by the local authorities, he does it well, and the apes are well cared for.

Now the average IQ of a bonobo is actually roughly measurable—bonobos are, by animal standards, very intelligent. Measured against a human scale, a bonobo's average IQ would be of about 5 or 6 (whereas the average human IQ is by definition 100.) Because the IQ scale is not linear, this does not mean that humans are twenty times cleverer than bonobos (by whatever yardstick of objective intelligence may be applied), but an IQ of 100 is rather something like perhaps ten thousand times or so more intelligent than an IQ of 5. In other words, your neighbor the veterinarian doctor is about ten thousand times or so more intelligent than his wards (the precise figure being of little relevance here, as long as it's clear that the man's IQ, although finite, is very much higher than that of the animals.)

Anyway this neighbor is a bit of an oddball. It turns out that he has been training the bonobos to ... somehow worship him. He has been training the bonobos to behave in a certain way in order to pay homage to him and his glory. When the bonobos neglect to conform to certain rites which the neighbor has devised that he deems reflect his glory, he harshly punishes the animals. He whips them, sometimes kills them. He has taught them to gather in rank and file, to wear certain pieces of cloth when he blows a certain whistle in the evenings, calling them to his worship.

Soon however the behavior of that neighbor came to the attention of the authorities. Word leaked, rumors spread, concerned neighbors espied his weird doings. Soon the authorities intervened; after due process, the good doctor was certified and put away. Once a well-regarded man who enjoyed respect within the community, the good doctor never recovered his reputation. His unexplainable need to be worshipped by lesser life forms was widely seen as totally detracting from his status as a wise and well-respected figure, and also ended his erstwhile renown as a capable veterinarian.

The only bit of mathematics in the above tale is that of the IQ points (10). It however lies at the root of how the situation unfolds and of the conclusions drawn: psychiatry, a scientific discipline wholly separate from mathematics, made a consequential numbers-based judgment that the neighbor was mad and put him away. If instead of bonobos, the wards had been human disciples with normal IQ's, the veterinarian would have deemed been a sect leader and not immediately nor necessarily mad. The proof that the man was mad is therefore a *category B* proof—proof buttressed by the IQ numbers—but not exclusively mathematical.

A key point is that, regardless of the exact IQ distance between the mad neighbor and his wards, loosely estimated at a factor of ten thousand or so, this distance is *finite*: the neighbor was only something like thousands of times more intelligent than his wards. He was not *infinitely* more intelligent than they were, yet his need for adoration on the part of lesser life forms clearly marked him as deranged. But by any acceptable definition of a Godhead, the Godhead's IQ is infinite. Mathematically, this means that the distance in intelligence between man and the Godhead is *infinite*, which in turn also means that this distance is infinitely greater than the finite distance between the mad neighbor and his apes. This throws off the question as to why and how an infinite Godhead would stand in any need of being worshipped by man, or of having us conform to any kinds of rites or strictures as to how we behave or dress or whatever, at least as long as we don't hurt anybody (11).

Two categories of considerations could conceivably modify the above conclusion.

The first possible objection is that intelligence is not necessarily the only or even the main criterion. Perhaps something equivalent to an EQ—emotional quotient—should be employed instead, or some other yardstick or combination of yardsticks. But this would not essentially alter the underlying argument: for instance, an infinitely good and compassionate Godhead would *also* have an infinite EQ. Animals are, like we are, capable of empathy. A Godhead capable of *infinite* empathy would dwarf both human and animal capacity for empathy, irrespective of any ascertainable distance between the latter two.

Emergence

A second and far more compelling consideration would be that at such very high values of IQ, EQ and any other attributes, the inescapable phenomenon of *emergence* would occur, and that, in environments where very large metrics or numbers or collections of anything intervene, there is absolutely no way *in principle* by which we can foretell or second-guess anything, such as behaviours, thoughts or events.

Emergence is the appearance of utterly new and inherently unpredictable properties and phenomena, ineluctably triggered whenever huge numbers, sizes, or any associated metrics start heaping up on top of some previous situation or status, and become involved in some way. Emergence results in novel, utterly unforeseeable phenomena, and often in profound qualitative shifts which cannot *in principle* be foreseen or anticipated in any way, and which show up—*emerge*—when the relevant associated metrics, numbers, or characteristics grow to immense proportions, or whenever the envelope of some applicable norm is pushed far beyond its usual moorings.

Quoting briefly from available literature (12), emergence is for instance the property that arises when one has *one million dollars* at one's disposal, as opposed to having *one million times a single dollar*. When equipped with merely a million times one single dollar without emergence, the most one could buy would be a million times cheap one-dollar (or less) items, such as, say, a million bubble gums. Emergence occurs when the dollars are allowed to pool together into great numbers, and new properties arise: say, a yacht can now be bought rather than a great number of cheap items. Thus, should you be given a million dollars with a stipulation that no emergence is allowed, you'd still be poor—constrained to one-dollar (or less) item purchases. Another example of emergence would be that of a viscous traffic jam, an emergent phenomenon arising from a large number of times the single phenomenon of one car being driven on a given stretch of road. In physics, odd quasiparticles emerge when vast collections of subatomic particles come together within constrained environments in space and time.

These emergent quasiparticles typically have strong, measurable effects on the collective properties and behavior of the group of particles.

Emergence can occur both within material contexts and within abstract or mental domains, as long as they build on scales significant enough to trigger the phenomenon. Knowledge is one such area: an early primary school student cannot begin to even conceptualize the contents of, say, an MIT Ph.D.-level course in engineering: the course builds on, and indeed *emerges from*, too many prior layers to be simply imaginable or in any way predictable from a much deeper layer.

One of the most striking examples of emergence is found in biology, in the odd case of the *Dictyostelium discoideum, aka* the slime mold. When conditions are good, the slime is made up of tens of thousands of independent, single-celled *fungus* spores. A change of environmental conditions can cause the many spores to congregate and become a single entity—an emergent single, coherent slug-like multi-cellular *animal,* which only appears if there are enough spores. In that case, as in the case of the dollars, emergence profoundly shifted the very nature of reality: out of a vast collection of loose independent spores arose one animal. The examples are endless.

Emergent phenomena keep building on rising numbers: the first ones appear at a certain level of magnitude, then further emergence keeps appearing when the resulting operative numbers are in turn vastly further multiplied, and so on. When the relevant numbers however become *infinite*, then whole new categories and *qualities* of emergent phenomena are apt to appear: even in the extremely rare (usually immaterial) cases where emergence somehow did not manage to arise within the finite ranges, irrespective of how big the relevant numbers were, the mere onset of infinity always triggers unexpected, novel emergent phenomena, which can then lead to familiar concepts losing their erstwhile well-established meanings. A straightforward example of this would be the simple operation of multiplication, as in 2 times 2 is four, as well as those of addition, subtraction, and division, when applied to infinities: the instant infinities become involved, these operations become something profoundly different from what they were prior. For instance, multiply any number no matter how large by zero, and the result of this multiplication firmly remains zero. Zero multiplied by a trillion zillion googols firmly remains equal to zero. But multiply zero by infinity, and the result may now take on any value between zero and infinity: the result has demonstrably become indeterminate (13). Or, multiply infinity by any non-zero finite number, such as one or two or one hundred or a few billion billions, and the result of this multiplication will stubbornly remain the very selfsame unchanged infinity you started with prior to multiplying, confirming anew that the very essence of the operation of multiplication has undergone a profound change.

This phenomenon of unpredictable emergence also constitutes conclusive proof that it is utterly *impossible* for mere finite mortals to second-guess a Godhead, *infinitely more so* than it is difficult for someone who has but a few dollars to imagine how billionaires live, or for a zoo animal to conceptualize the contents of a Ph.D. thesis. *Unless,* that is, *we are ourselves somehow infinite*—ourselves a Godhead. Even so, we could likely only successfully second-guess a Godhead in such areas where we would still be, somehow, demonstrably infinite—thus probably not in, say, intelligence, where everyday experience conclusively proves that our IQ remains firmly in the finite range.

There we have it: proof positive that it is irretrievably beyond the ken of mere mortals to ever *second-guess* Godheads if such exist, and to ever presume to speak in their stead or on their behalf. If anyone says that they know how a Godhead thinks and what It wants, and that they are therefore entitled to speak on Its behalf, they are anthropomorphizing the infinite, and the reality is that they cannot even *in principle* begin to understand infiniteness. They cannot possibly have a clue.

Some spiritual traditions and/or cults tell of soothsayers, of privileged spokespersons somehow enabled and entitled to speak on behalf of an (inevitably disincarnate) Godhead. Perhaps some among these soothsayers are themselves no mere mortals at all, but incarnate Godheads, temporarily seconded here to teach. As it is, the inherent inability to ever second-guess a Godhead, or to faithfully relay Its thoughts or speak on Its behalf, demonstrably also fully applies to revelation and revealed truths, irrespective of whether such would be conveyed by mortals or by temporarily incarnate Godheads.

Here's why.

The reverse of emergence, which we'll call *demergence*, is what happens either by simple scaling down (say you lose 999,999 dollars from your erstwhile million dollars and are left with a mere lone dollar), or through some other means (for instance, you keep your million dollars but are now forbidden to pool them and may only spend them one by one), or, say, you used to live in the three dimensions of normal three-dimensional (3-D) space but have now somehow become constrained to living in 2-D space, like a lichen on the face of a tree or rock. As we shall see, most of the wealth of phenomena and benefits that arose from emergence are irretrievably lost when demergence occurs, and can only return if and when emergence happens again—when the large numbers, or infinity itself, return.

A telling illustration of demergence is seen in mapmaking, where a representation of our spherical, three-dimensional Earth must be transcribed onto a flat, two-dimensional medium to serve as an ordinary map. Because one dimension is lost in the transcription, demergence kicks in and warping, garbling, and irretrievable loss of information inevitably happens. As a case in point, the most widely used 3-D (three-dimensional) to 2-D (two-dimensional) map projection in the world, the so-called Mercator projection, totally distorts the resulting 2-D maps: for instance, Greenland appears to be about twice the size of Australia on a standard Mercator world map, whereas in reality Australia's surface area is about four times bigger than Greenland's. Alaska appears to be roughly the same size as the continental United States, whereas it actually is about five times smaller. There exist other map projection schemes, such as the Winkel-Tripel projection, or the Peters projection, all of which are plagued by the inescapable fact that for instance the poles, point-like and dimensionless in the three-dimensional reality of Earth, must be represented in 2-D by the full top and bottom sides of a 2-D map.

Information loss and data distortion multiply very quickly when the loss of further dimensions occurs, in other words with further dimensional distance from the lower-D canvas where we attempt to map, transpose and understand a higher-D reality. For instance, whereas a four-dimensional *tesseract* hypercube still remains dimly recognizable and somehow comprehensible when drawn onto a sheet of paper, any attempt to draw onto a 2-D sheet the representation or projection of an even higher dimensional figure, such as a nine-dimensional *noneract* or any other such hypercube, leads to immense distortion and garbling, and to unmanageable information loss. This loss is twofold: on the one hand, there is the data warpage and distortion that we remain cognizant of (such as the point-like poles of a 3-D sphere that we know have become line segments under a 2-D representation.) But there are also information or data losses that we are unaware of, because dimensional downscaling has led either to the outright disappearance of even the traces or vestiges of such data, or to their seamless conflation with separate data from other origin (for instance, if we were to project a Mercator map onto a one dimensional line segment, we would become totally unable to separate out the now seamlessly merged borders of the different continents within the line segment.)

Distortion and information loss are sometimes called rendition errors, or filtering errors. Such filtering errors would become *infinite* if an infinite dimensional reality were mapped onto a finite-dimensional environment. Any Godhead would most likely be, to use a Freeman Dyson phrase, *'infinite in all directions'*, at least inasmuch as described mathematically, and as such would have to be infinitely multidimensional (14).

Now let us assume that there *is* a Godhead and that there are such things as genuine mystical experiences. Let us assume in the first instance that the experiencer is a mere human. As will be mathematically demonstrated further on, any mystical experience *cannot* be exclusively left-brain analytical, but must concurrently occur through revelation, or some other transcendent means. How the experience will be interpreted and remembered—brought back to Earth, as it were—will inescapably depend on the constituent characteristics of the particular individual experiencing the revelation. It will be differently *filtered* by each experiencer's finite make-up, in a way vaguely reminiscent of how differently 2-D map projections, the Mercator and the Peters and all the other projection schemes, reflect, render and report an essentially alien, because higher-dimensional, 3-D reality.

The ensuing unavoidable discrepancies and variations in the way different people filter into their everyday environment their mystical experiences explain in one fell swoop why, even if such experiences are real and do in fact spring from some outside reality, *revealed cults routinely contradict one another.* It also strongly suggests that, at the very least, all revealed cults cannot be the complete tale—both because of the experiencers' built-in inability to grasp fully the essence of something that is infinitely emergent above where they are, and the unavoidable fact of their filtering infinity into some finite terms of reference, thereby irretrievably and unrecognizably warping it (15).

Some have argued that because mystical experiences transcend material limitations, they can be transcribed forward into material reality without being distorted. It does not work, on compelling theoretical and experimental grounds:

- Experimentally, the conflicting and wholly incompatible dogmas that have historically sprung up from the visions of different experiencers who went on to establish various cults and churches—the Joseph Smiths, Hong Xiuquans, et al., of the world—show how even broadly similar mystical experiences can lead to totally different interpretations, underscoring the impossibility of faithfully transcribing mystical experiences into everyday reality. (For the sake of exhausting all of the theoretically possible alternatives, let's mention the two other possible alternative explanations that either 1- all of these experiencers are liars, conmen or madmen, a view that remains unproven the face of historical evidence, or 2- that the experiencers have not met with the same Godhead and hence the experiences are different—which however immediately leads to a self-contradiction since all revealed cults are monotheistic and do not admit of other Godheads.)

- In theoretical terms, this view cannot make sense either. It would be akin to saying that, say, a wonderful symphony, *for the sole reason* of its sheer inherent quality, could be rendered without information loss and corruption on lesser audio equipment with poor dynamic range, or that a superb TV show in magnificent colors would be seen in full on a black and white set with a cracked screen, that a map would be true *if* the world it represents were wonderful, and so on.

But what about cases of allegedly *incarnate* Godheads? Can a spokesperson, should that spokesperson happens to Itself be an incarnated Godhead, possibly and legitimately speak on behalf of a greater, disincarnate Godhead, and relay Its thoughts faithfully? Some traditions do indeed speak of a Godhead incarnating, at least in part. For instance, Christ is said to have been a godlike part of Godhead who 'became flesh' (which in turn gave rise to a whole figurative vocabulary, with for instance the still disincarnate part of Godhead called 'father' and Its incarnated part the 'son'.)

The built-in limitations of 3-D life would kick in as well in the case of an incarnate Godhead—there simply does not exist any mechanism whereby the curbs and curtailments inherent to the very status of incarnateness can disappear from within that status. Certainly, an incarnate Godhead would presumably be quite able to switch back and forth between an incarnate and a disincarnate state at will, but the instant It would be back in the limited, 3-D, material world—the world of the flesh—It would necessarily become enmeshed in its attendant limitations and inescapable filterings—if not, It would instantly become unmoored. Emphasizing the point, there is mathematical proof, as we shall see further on, that if a Godhead exists, It must be omnipresent in the multiverse. The instant a Godhead incarnates, the proof breaks down, and omnipresence immediately vanishes, along with any number of other godlike attributes.

Still other traditions, rather than singling out a few special people from the rest of mankind, view *every person* on Earth as an incarnated divine spark off a larger Godhead. In which case it all becomes even simpler, as everyday experience plainly shows that unforgiving *demergence* has kicked in, and there is precious little trace of whichever infinite abilities and godlike traits we may have enjoyed before reducing to a demergent, incarnate state: our possibly divine nature becomes occulted and beyond our reach the instant we incarnate.

Finally ending the credibility of self-styled spokespersons presuming to speak on behalf of Godhead, it turns out that, even if someone were against all impossibility still somehow privy to Its thoughts, It could not possibly convey these thoughts by means of even an *ideal* human language. Human language cannot function as a viable vehicle for godlike thoughts.

Here's why.

Irrespective of word order, be it subject-verb-object as in English, or subject-object-verb as in Japanese, or any other such pattern, thoughts and qualifiers must be expressed *sequentially* by language, one at a time, because we can only speak and write one word at a time. Correspondingly, language is written linearly, either left to right, or right to left, or even in columns top to bottom or bottom to top, depending on the particular language. Because human language must thus proceed linearly (i.e., in 1-D), the selfsame warping and loss of information that besets mapping also fully applies to language, and effectively bars conveying thoughts from higher dimensions into human language.

To illustrate this limitation, imagine that you'd want to write a simple essay on spacetime. You would perhaps start by entitling your essay 'Spacetime', but you'd immediately find this title unsatisfactory, because you would have unduly positioned the word element *'space'* before the equivalent element *'time'*, and what you are attempting to do is precisely stress the equivalence and full interchangeability of the two concepts of space and time (16). You then try to shift around the order of the word elements and come up with Timespace—every bit as unsatisfactory, for the selfsame reason. You then try capitalizing the initial letter of every word element, and you end up with either SpaceTime, or TimeSpace— it's better, but not by much. There is still this stubborn, unwelcome order of appearance of the two concepts, imposed by the 1-D linearity of language, inescapably leading to a sequential ordering of the two concepts, which you are striving to eliminate.

Then you hit upon the solution: you write the word SpaceTime in an unbroken looping circle, in the shape of an ouroboros (a snake biting its own tail.) The two constituent word elements have now become equivalent, embedded within the lettered closed loop where neither word element is first nor second. You have solved an intractable issue by, in effect, upping to two-dimensional writing, occupying a paper expanse both lengthwise and heightwise—a higher-dimensional text than the usual 1-D script allowed by traditional writing. This extra dimension is the only way you can faithfully render your meaning. Should you now attempt to read your title to someone who can't see it as written, for example over the phone, you can no longer do so easily: because you have moved from the usual 1-D linear expanse of text to a 2-D planar expanse, and since there is just no way you can speak in 2-D, you now must fully *describe* it rather than simply read it to convey the full meaning (in mathematical terms, this added-on description would be called *meta*-information, or meta-data.) Unless you do so, reading it will result in loss of key information—because of the demergence between the two dimensions of your text and the one dimension of language (17).

- Then, in the course of drafting your essay, difficulties mount. You find that before you can convey to the reader a certain concept A, you must lay the groundwork for its comprehension by first imparting to your readers another concept, B, which happens to be indispensable to understanding A. But B in turn depends on an array of other concepts, which themselves can only be understood within the framework and environment of yet another set of other concepts, elements of which *will only appear relevant and interesting if one already appreciates A itself.* So in the end you need to understand B before you can understand A, but before you can understand B you must comprehend a whole kindred context—which only appears interesting and relevant if one already appreciates the essence of A itself. Although no *causal* circularity is involved, there is a clear cross-wise co-dependency which can make, in the absence of an appreciation of A, any preparatory build-up work seem irrelevant, pointless, unilluminative and boring: the very act of comprehension requires a holistic, multi-dimensional approach involving a number of simultaneous feedback loops and cross-wise relationships of meaning. And then you discover more such transverse loops. And there you are, stuck in your writing because of the limitations of sequential 1-D human language. Any text you will manage to write will thus *in principle* be imperfect—if will of necessity be a compromise, and thereby inevitably open to a measure of misinterpretation, incomprehension, and carping (this scenario is, incidentally, by no means a mere academic view, but something that routinely happens in philosophical communications.)

In an *ideal* human language, the lexicon would be much bigger and there would exist a separate, independent word for spacetime. However, we only ever converse in actual languages, not in hypothetical ideal ones; worse still, an ideal human language would still demonstrably *not* solve the issue (18).

From a more literary perspective, how do you describe with words color to a blind man? Eric-Emmanuel Schmitt, a well-known wordsmith who holds France's highest academic degrees in literature and is the author of several best-selling novels in Europe, wrote in 2015 a best-selling memoir of having, decades earlier, undergone a transformative mystical experience when he lost his way in the Hoggar mountains of Algeria. He writes: "Words, pallid words, cannot begin to describe what I'm experiencing. They've been designed but to describe stones, things, feelings even, mere human or near-human realities. How could they ever describe what is so irretrievably beyond them? How could our limited words describe the infinite? How could the labels of the palpable ever be stamped upon the ineffable? Their job is to inventory the world, they're rooted in the soil, but I'm soaring so far away. No longer do I think in coarse sentences, no longer does my experience come from eyes, ears, or skin." The experience turned him from his earlier staunch atheistic stance to a new, personal form of agnosticism.

The impossibility for any one language to demonstrate the truth of any statement expressed within or by that language, is even demonstrated mathematically. The proof is called Tarski's undefinability theorem, and in essence it says that you cannot prove the truth of a statement such as 'this car is blue' from within the language itself—that outside corroboration, not narrowly involving the language only, is needed.

Religious texts—scriptures—full of pearls of wisdom on the one hand, often however happen to contain quite surprisingly objectively horrid and/or objectionable passages. Jonathan Sacks, in his 2015 book 'Not in God's Name', tries to make sense of the worst instances of such passages. He writes "These texts require the most careful interpretation if they are not to do great harm.... One who translates a verse literally is a liar: no text without interpretation; no interpretation without traditions.... Religions have wrestled throughout their histories with the meanings of their scriptures, developing in the process elaborate hermeneutic and jurisprudential systems. Medieval Christianity had four levels of interpretation: literal, allegorical, moral and eschatological..... *etc. etc.*" It is both the power and the beauty of mathematics that no interpretation, no further 'interpretation of interpretations' are ever needed, there is no need to jump through hoops and hoops of probably controversial and not universally accepted ifs and but's and maybe's to make sense of the texts. A textbook of mathematics at whatever level is to be taken at face value, and whenever doubt arises it does so on sound mathematical and logical grounds, and the envelope of possible interpretations is both clearly delineated, and signposted by the math itself. The ultimate weakness of Sacks's and similar arguments, is that they squarely violate Tarski's undefinability theorem, as they try to make sense of controversial statements from within the same system of logic which these statements sprang from in the first place: it can't be done. As we saw earlier, should we attempt any other scientific approach than that of fully neutral mathematics, we risk running into the 'butterfly effect' issues that subtle cognitive bias can have on analysis outcomes.

If we broaden the definition of language to include non-verbal languages, such as gestures (such as so-called *mudras* or any other gesticulation) and music, we still hit upon limitations, although of a different nature. Whether or not there exists an infinity of possible gestures or mudras depends, again, on whether space is continuous or not: if space is continuous, there is an infinity of possible, separate gestures. Not so if space is granular: if, against all the signs pointing to the contrary, space is actually continuous and hence there exists an infinity of possible mudras, that infinity would still be low-ranking in the infinite hierarchy of infinities, and still infinitely unable to express the full complement of thoughts stemming from a higher-ranking infinity (we shall soon visit this issue of multiple infinities.) As far as music goes, the situation is simpler, because we can broaden the definition of a 'note' to encompass any arbitrary pitch (i.e. the number of vibrations or *cycles* per second of a note: the note *A* for instance has a pitch of 444 hertz) to any arbitrarily small degree of difference within a continuous range.

We can thus define any bounded continuous range of pitches as a language (with boundaries set at two end points respectively marking the lowest and the highest available pitch, with an infinity of different sounds in-between), and define some specific meaning attached to any sound within the range—say, some meaning the note vibrating at exactly 444.0011 cycles per second, with another meaning associated with the note at, say, 444.00111 hertz, and so on) and hence theoretically we would be able to convey an infinity of separate meanings (the issues of whether we'd be able to differentiate between close pitches being a separate issue.) Music therefore appears to be the ideal language to convey the widest possible range of thoughts in our limited, lower-D environment—a potentially infinite range. But because we cannot possibly define consensually an infinity of meanings within any given range in finite time, the issue now becomes one of *interpretation*: we have come close to solving the issue of the limitedness of language, at the cost however of falling straight back into the issue of rendition we encountered earlier—if the ideal language of music is used, then we will inevitably risk interpreting what is said in different, sometimes mutually incompatible ways. The limitations of our world thus wind up always popping up one way or another, and conspire to garble any possible communication with higher-D environments—with the higher realms if such exist.

To compound it all, there is also, again, the further issue that the infinity of the pitches available within a range of sound vibrations is also a lower-ranking infinity—technically one of aleph-one size (rather surprisingly, infinity comes in different sizes, so-called *cardinalities*. The aleph scale of the measure of infinities shall be examined further on), and the ideal language of music therefore cannot in principle be used to describe elements stemming from worlds of a higher cardinality than aleph-one, i.e. the very realms where a Godhead would presumably dwell.

To communicate effectively all the nuances and information loops present in complex, interactive or interdependent thoughts, we'd need a parallel, multi-channel communication facility, rather than the single-channel mode available through oral and written language. In other words, we would need a matrix-like, web-like, multidimensional language (18), a way to execute wholesale memory dumps—the equivalent of instant multi-channel communication. Human language does not, and can never, permit that option. One attempted way of alleviating these limitations to some extent has consisted, in some cultures, in broad-based holistic approaches to the business of life—which in turn has led to other issues. Other cultures still have favored reductionism over holism, i.e. the analytical breaking down and parsing of reality into independent components, an approach which in turn has led to a different set of issues (19). These different approaches are reminiscent of the way various map projection schemes are used. Ultimately, they are all in varying degrees inadequate, little more than band-aids trying to paper over the intractable inadequacy of our available communication modes whenever we try to convey higher-complexity or higher-dimensional realities.

Inescapably, trying to communicate the ungarbled thoughts of an infinite Godhead through human language would prove infinitely more impracticable than properly phrasing a philosophical essay or adequately conveying its title. Any attempted rendition would *in principle* turn out so inadequate as to be useless. Even if someone were, impossibly enough, privy to the thoughts of a Godhead and able to understand them, any bid to couch such thoughts adequately, let alone faithfully, by means of any possible human language would be doomed to failure. Whichever way we may turn, dogmas and alleged revealed truths do not look tenable. Whenever a televangelist or anybody else tells us again that they spoke with God last night (20), they are either lying or just being delusional.

The fact that any human language can only be one-word-at-a-time sequential and hence one-dimensional also ends the eventuality that any written texts could ever have been composed by a Godhead: if a text is written in any human language, even an ideal one, then either it was written by humans, or alternatively it was written by a Godhead in a state of such infinitely deep demergence that any godlike content was irremediably lost. If anyone, or for that matter, any written text says that on account of some Godhead you must get circumcised, wear this or that piece of clothing, hurt someone or whatever else, the hard fact is that it simply cannot possibly be the case. Equipped with our finite IQ's, confined to life in a 3-D environment, we cannot possibly even remotely venture an approximate guesstimate, let alone understanding, of how any possible Godhead would ever think or what It ever could possibly want. In the context of Infinity, the very words 'think' and 'want' themselves have inevitably acquired new emergent meanings that we are not equipped to imagine nor conceptualize. What we certainly can do, however, is some math and thereby, paraphrasing the aviator John Gillespie Magee, slip the surly bonds of Earth, put out our hands, and somewhat approach the face of Infinity.

It is worth noting that any given believer who whole-heartedly adheres to a particular cult or set of dogmas has no choice but to deem the whole assortment of *other* dogmas, from competing and incompatible faiths other than his or her own, as plain wrong—*not* valid truths. Separate religions and cults out there are mostly mutually incompatible—which is the reason why they exist separately in the fist place—and therefore the only possible alternatives are that *either,* if one particular set of dogmas actually happens to be true, all of the other incompatible dogmas are not revealed truths, and hence bear no relationship to divinity and are wholly man-made—*or* alternatively, that *all* dogmas are man-made. A straightforward calculation, using the respective current numbers of the followers of diverse cults and religions across the globe today, shows that in any scenario, *most* religious followers in the world today follow sets of dogmas that cannot possibly be true and are therefore man-made.

Demergence, and its resulting warpage of higher dimensional realities, may also help solve a conundrum of time. In a nutshell, the issue of time relates to the assumption, borne out by strong evidence in favor of a Big Bang, that our universe had a beginning. But it has been argued elsewhere (21) that time itself is a demergent feature which only appears in universes of lower dimensionality, and that in higher dimensional universes or metaverses the very concept of time becomes different. Therefore, the very concept of a 'beginning' of time may lose its meaning (even in lower-dimensionality universes, there may exist areas of spacetime made up of closed timelike loops, something akin to Möbius strips in time, where no beginning could exist.) Under this view, our universe had a beginning only because it is lower-D universe, perhaps a small provincial corner of a much wider metaverse which includes timeless, higher-D realities. In one fell swoop, all the chains of reasoning predicated on the beginning of our universe are swept away and subsumed into a picture of a possibly much wider reality.

Defining Godhood

Before we are able to go on, we must attempt a simple and perhaps provisional definition of a Godhead, as consensual as possible. A or *the* Godhead must by most definitions have some infinite attributes, traits or characteristics—it is supposed to be, indeed *defined* as being, infinitely good, infinitely intelligent, infinitely powerful ('almighty'), infinitely knowing ('all-knowing', *aka* 'omniscient'), and so on, notwithstanding the few minority theologians who dispute that a Godhead is infinite or almighty in any way, with whom we have disagreed on a variety of grounds (most compellingly because we deem the presence of infinity to be the disruptive factor necessary to beget the very essence of Godhood in the first place.)

In a bid to start out with a broadly consensual definition of Godhead, we will define a Godhead as some entity who is:

- 1- Alive,

    and

- 2- Who possesses at least one, and possibly several traits or characteristics that are demonstrably infinite (for instance, who is infinitely intelligent, or good, or powerful, or lives in an infinity of universes, or an infinity of dimensions, or all of the above, and so on. Of course, as we shall see the presence of even one infinite quality may well result in the presence of more.)

There exist also other, perfectly valid alternative definitions of a Godhead, although these may become a bit more elaborate. For instance, a Godhead could absolutely be defined as someone who knows, or is able to know, both the speed and the location of any moving object at the same time with any arbitrary level of precision. That's it. Simple as that. We mere humans are prevented from being able to know simultaneously, beyond a certain level of accuracy, the value of certain pairs of attributes attached to physical objects (such as the speed and location of a moving object, although the moving object most definitely *has* a given speed and *is* at any instant in time at some particular spot in space). The physical law that curbs the amount of simultaneous knowledge that we can ever achieve is called the Heisenberg uncertainty principle. Anyone capable of overcoming the limitations of any incontrovertibly proven physical law such as Heisenberg's principle would be a Godhead under any definition (we will revisit this theme more in-depth later on.)

Objections to the definition adopted above can legitimately already be raised. One first immediate possible such objection would be that infinity itself may actually not *exist* at all in actual reality—the concept of a materially existing infinity might just be a mistaken thought and an illusion, having no material counterpart in any actualized reality anywhere. Concluding however that a Godhead does not exist if infinities do not exist in our reality would be circular reasoning: it would be saying, in effect, that the Godhead does not exist because It does not exist. Should infinity however demonstrably not exist in *any* reality, it would leave us with the two possible alternatives that either Godhead itself does not exist in actual reality, or that It is finite. We will examine this question, and reach a mathematical conclusion that whereas there theoretically is room for infinities to exist in possible, logically conceivable realities, ascertaining their factual presence proves more problematic.

Many Infinities

The next immediate objection is that, if they do exist, there are, as was mentioned earlier, very many different infinities—in fact, broadly, an infinity of infinities (22). Infinity can take on different *infinite sizes* (often called cardinalities, or strengths). Much less surprisingly, infinities can be of different *qualities* (the quality of an infinity is what is contained within that infinity.)

- For instance, the endless series of whole numbers 1, 2, 3, 4, 5 .... could be augmented by sprinkling in a few halves, for instance as follows: 1, 2, 3, 3.5, 4, 5, 5.5, 6 ..... These insertions would demonstrably not alter the size (*cardinality*) of that infinity. The endless series of whole numbers with merely, say, the number one and a half added into it would however be already *qualitatively* different from the original series, although it is easy to conceptualize that the size (strength) of its infinity would not be altered from what it was before.

- We could even add in an *infinity* of halves—perhaps every tenth half or third half or second half or even *all* of the halves there is—it still would, somewhat more surprisingly but demonstrably, not alter the cardinality of the series (23). The *quality* of the infinite series, however, would be different in each case.

Besides these two attributes of cardinality and quality, infinities have a third property which may take on different values, that of *boundedness* (a property which will play a role here): infinities can be *bounded*, i.e. contained within set limits (for instance, there exists a full infinity of real numbers between the numbers 0 and 1, wholly constrained within the limits set by the two end numbers 0 and 1), or *unbounded* (e.g., there exists an infinity of numbers between minus infinity (*written* -∞) and infinity (+∞). An infinity can also be partly bounded (keeping to the same example, that would be the infinity of numbers that lie between, say, the number 5 and infinity.)

To take an example from the physical world, time inside a black hole can exhibit infinities—time towards the past would be seen from within the black hole as extending to infinity, but from outside it would be seen as bounded—its end point would be the instant when the black hole formed. Equivalently, any time inside the black hole would lie beyond the infinite future of time outside the black hole—they would be instants of time for ever unattainable from outside the black hole.

The realization that there are different sizes of infinity, and their apprehension (and hence subsequently their definition) by means of a metric of infinity called the Aleph numbers, led Georg Cantor, the mathematician who first formalized the study of infinity in mathematics, to happen upon a *workable, purely mathematical definition of a Godhead.* His definition of a or the Godhead is extraordinarily simple, as follows.

Infinite series such as the never-ending series of whole numbers 1,2, 3, 4 ... *ad infinitum* are called a collection, or a *set.* In quite general terms, different cardinalities are present whenever some infinite set cannot be matched one-on-one to some other infinite set: should we try to establish one-to-one pairing between the constituent elements of a given infinite set with those of another infinite set, and fail, and an infinite number of unmatchable leftover elements remain. When that happens the cardinality of the latter set is then said to be bigger—or *'stronger'*—than that of the former. (An example of this is the set of whole numbers, which cannot be matched one on one with the set of all real numbers (i.e., numbers containing all possible decimal extensions.) The former cardinality is aleph-naught, whereas the latter is aleph-one, infinitely larger than aleph-naught, by a factor equal to aleph-one itself.

Now let us define a *subset* of the infinite series of whole numbers 1,2,3,4,5.... which we'll call N: a subset is any either finite or infinite collection containing within itself any number of elements taken from within N: for instance, (1,2) or (9,18, 20, 21, 47, 1200983683), or (8, 121, 1001), and so on, are all subsets of N.

Georg Cantor demonstrated that if you build a set (which we'll dub the power set of N, written "PS-N") made up of *all the possible subsets* of the infinite set N, then this new set PS-N has a cardinality bigger than that of N: from the original infinite size of N, aleph-naught, we have built a brand-new set whose size is now aleph-1, infinitely stronger (bigger) than the aleph-naught size of the N set we started with. In other words, the cardinality of the set of all the subsets built from any given infinite set goes up by one tick over the cardinality of the original set itself.

Cantor showed that this is a general result that applies to all infinite sets: whatever an Infinite Set A may be, PS-A is always of a higher cardinality than A, which also means that any individual subset from within A is always smaller than PS-A (24). This also means that we can build ever new and ever stronger sets, through simple recurrence, from any base infinite set: for instance, the power set of the power set of N is stronger than the power set of N, and its cardinality is aleph-2.

The process can be repeated endlessly, leading to ever bigger cardinalities, on the way to an elusive, never-attainable aleph-infinity.

This is where, according to Cantor, God Itself inevitably enters the picture: let us look at the set of all possible sets in the universe or multiverse, a set which contains all possible sets everywhere and everywhen. We will call this set Ananta, from the Sanskrit word *ananta* meaning infinite (25). The cardinality of the Ananta set is thus, by definition, aleph-infinity.

By definition, this new Ananta set, being the set of all possible sets within the universe or metaverse (whichever is applicable), has no choice but to contain PS-Ananta itself (the power set of Ananta), which leads straight into a mathematically inextricable situation: on the one hand Ananta is strictly (26) smaller than PS-Ananta, for all sets are always strictly smaller (*weaker*) than their power sets, but it is also definitely strictly bigger (*stronger*), because PS-Ananta is an individual subset of Ananta, and subsets are always strictly smaller than the set itself.

This is a stark, blatant core contradiction, which lies at the very heart of the mathematical approach to infinity—an unresolvable *antinomy*, in Georg Cantor's word.

Georg Cantor resolved the issue with a neat sleight of hand: he stated that Ananta cannot be possibly understood by logical analysis. Because of its sheer size, the wonted norms and tools of analysis, be they mathematical, quantitative, discursive, or rational, all break down and fail—the ultimate case of emergence, where all definitions, all understanding, and mathematics itself have fully broken down in the face of aleph-infinity cardinality.

This Ananta set, he claimed, is the Godhead Itself—the creative source of all that exists, all that can ever be, and it can only ever be approached by intuitive insight, however remotely (thereby providing a mathematically sound if narrow justification for Einstein's well-known later aphorism that 'imagination is more important than knowledge'.) With this, Georg Cantor came closer to a possible mathematical definition of Godhead than all the ontological thinkers ever did (27).

In Search of the Infinite

Where can infinities be found? Conceivably, they could nestle both within material reality—the reality of things and of space and time—and abstract reality, that of thoughts and mindscapes and numbers and art, and in unseen worlds.

In *material* reality, infinity could in principle lodge either in the infinitely big, if the universe or metaverse happens to be of infinite spacetime expansion (and our visible universe but a small part of a larger, infinite universe or metaverse), and it could also nestle within the infinitely small, if our space is endlessly divisible, or in both, should the universe/metaverse be both of infinite extent *and* our space or spacetime infinitely divisible (28).

As it is, it is highly unlikely that space is infinitely divisible, and there are compelling grounds to believe that it is made up of finite, indivisible 'atoms of space' instead. According to credible recent calculations, these 'atoms of space', also called 'quanta of space', are extremely small—there would exist about one followed by one hundred zeros of them inside every single cubic centimeter of space—but that number is nevertheless finite, and there is therefore probably no such thing as infinitely small (29). Of course, at the other end of the scale, beyond our known universe, the wider universe may well be of infinite expanse. It could also be infinite in other, more exotic ways.

But whether or not material reality can accommodate infinities turns out to be largely irrelevant here: regardless of whether spacetime is infinite or not at either end of its expanse scale, the infinity of spacetime *cannot possibly accommodate an infinity of aleph-infinity size*, because even if it exists, spacetime infinity is necessarily firmly constrained to the lower end of the scale of cardinalities. The cardinality of plain infinite space is aleph-1, even in a hyperspace of any number of dimensions. There exist contrived ways whereby the cardinality of the infinities capable of being lodged within an aleph-1 infinite spacetime could be lifted by a tick or two, but not by much further.

What, then, of *abstract* reality, for instance thoughts? Although thoughts may not be graspable in the same sense as, say, chocolate boxes or houses, they are nevertheless every bit as real (30). Can there be an infinity of them? We are touching here upon the very definition of what legitimately constitutes reality, an issue that we will revisit in Part 3, but for the time being suffice it to say that, whereas it is easy to merely vaguely or formlessly *conceive* of infinity (all we have to do to do so is to, for example, think of an endless series of numbers, such as all the possible decimal numbers between 1 and 2, or simply all the whole numbers) it is equally plain that only a finite quantity of these numbers has ever been, and shall ever be, *specifically*, narrowly, singly envisioned within any thought by anyone, anywhere, any time.

For instance—whereas, should we ever think about it, we'd be very well aware that the following number belongs to the infinite series of decimal numbers between 1 and 2:

1.0007736544267399265246789999326763562580988162564 15 52537013,

this number has however *not* become embedded in manifest reality, under any definition of what reality is, unless and until it appeared individually either in some real-life computation, or unless and until someone thought of it *individually*, i.e. had their brain neurons firing in a way that brought up that specific number onto the radar screen of their awareness, and thus at some point in time imported it into the realm of existing thoughts—and by doing so, actualized that number into actual, manifest existence (31).

- After this number has been irrevocably ushered into awareness by focused thought, we may well go ahead and forget it: it has now irrevocably popped up on the radar screen of reality through real, actualized thought at some point in time, and shall remain forever actualized, for it did make an appearance at some point within the actual reality of spacetime. Before, it had never.

Somewhat unfortunately for us, a finite number called the *Bekenstein bound* sets a finite limit on the amount of information that can ever be embedded within a finite region of space in a material universe (32). As such, it places an upper limit on the number of thoughts that can possibly be held within any non-infinite volume of space at any one time, therefore on the number of thoughts that can possibly be held within a brain. (The *origin* of thoughts, i.e. the question of whether our thoughts are locally generated inside the brain or whether they somehow take their source elsewhere, with the brain then only serving as a kind of relay to intermediate them, is irrelevant here (33). For an average-sized human brain, the upper Bekenstein limit would amount to about 10 to the power 10 to the power 42—an unfathomably, hugely vast number, yet a *finite* number. What the Bekenstein bound thus says is that there is a limit to the amount of information that can be either generated or intermediated into our material world by a finite-size brain at any one time—and that *even if the source of the brain's information were infinite, only a finite part of it can ever be midwived into materiality within finite time.*

All of the above considerations have no bearing on the *existential* question, on whether a Godhead exists or not. They however demonstrate that if It does, It is beyond our ability to comprehend, and infinitely beyond our ability to ever second-guess. In one fell swoop, all of the cults and religions of man which purport to second-guess and know Godhead are shown to be largely illusory. But, as we will now see, they still mostly get it right, mathematically speaking, when they tell of a Godhead's attributes.

The Attributes of Godhood

A good place to start is to analyze three attributes widely seen as key attributes of Godhead: all-knowingness, omnipresence, and almightiness.

All-knowingness

Although we do insist on the existence of infinity for the narrow purpose of defining a Godhead, *many* attributes of Godhead do not necessarily require the existence of infinity, and could equally well function without it. As a case in point, *all-knowingness*: there are two possible definitions of 'all-knowing', both possibly equally valid.

1- The first definition would be 'infinitely knowing'. Should we adopt this definition, it would then stand in immediate mathematical contradiction with a finite universe, i.e. a universe where there is only a limited, finite (although huge) number of things than can ever be known. If this definition stands, then wherever any Godhead dwells must necessarily be a place where an infinity of things and events exists: this place must be either an infinite universe or an infinite multiverse.

2- The second possible definition would be 'knows all there is to know', even if all there is to know is limited and *finite*. All-knowing would then mean a Godhead who knows everything there is to know in the universe, but nothing more—on the solid grounds that there is nothing more to know.

Both definitions would work, although of course the former one is the only one compatible with an infinite Godhead.

There exists an A-type mathematical proof that *if* a Godhead is all-knowing, irrespective of whether It is 'finite' or 'infinite', then you and everybody else is then part of that Godhead.

Therefore, should someone tell you that, on the one hand, there does exist an all-knowing Godhead, but that, on the other hand, you yourself are not divine, that you are somehow separate from and not a constituent part of the Godhead and its very essence of Godhead, then that someone is saying the somewhat slapstick mathematical equivalent of one and one equals, say, fifteen and a half or a conga line.

Here is one proof (34).

- 1- An all-knowing Godhead is by definition capable of correctly answering any *answerable* question at all (bearing in mind that some questions, such as 'how long is a piece of string', are inherently not answerable.)

- 2- You, a mere mortal, now write out the following sentence: "*The Godhead will not say that this sentence is true.*" Call this sentence S. S can be equivalently rephrased as: "The Godhead will never say that S is true."

- 3- Now ask the Godhead whether S is true or not.

- 4- If the Godhead says S is true, then S ("the Godhead will never say that S is true") is actually false. Therefore the Godhead will never say that S is true.

- 5- If the Godhead says instead that S is false, it would immediately be in contradiction with itself since this statement then means that the Godhead says that it will say that S is true— not false as it just did.

Therefore: the Godhead will never say that S is true, *even though S is factually true* and *everybody knows it full well.*

Here's reality then: the all-knowing Godhead will never be able to call or confirm this truth, a truth however which you and any other mere mortal know full well is true and are able to readily confirm is true.

It's not simply a matter of the Godhead knowing the answer but somehow deciding to keep the truth to Itself either. To appreciate this, ask the Godhead: 'do you know the answer?' Whether the Godhead answers yes or no, we fall straight back into the above paradox: as far as the Godhead (and *only the Godhead*) is concerned, the question is equivalent to 'how long is a piece of string': there is simply no answer. In other words we positively know a clear-cut truth which the Godhead can never call. The Godhead therefore does not appear to really know everything that is knowable, and is thus not really all-knowing.

Unless ......

Unless the Godhead is within you as well, and knows the answer directly from inside you.

This constitutes a simple mathematical, type-A proof that *if* It exists, an omniscient Godhead must be within each and everyone of us—because each and everyone of us, if we only look, knows the answer to the question above. The above contradiction disappears if and only if the Godhead is inside you as well and is able to peek at the answer from within.

Other logical conundrums have been used to try and demonstrate that *in principle*, all-knowingness cannot be achieved under any circumstances. In his book 'Impossibility: The Limits of Science and the Science of Limits', John Barrow uses the example of the sentence *'Nobody anywhere knows that this sentence is true'* to probe the limits of omniscience. If the sentence is true, then a Godhead does not know that the sentence is true, although it plainly is, and hence It cannot be all-knowing. If it is wrong, then in turn this means that somebody somewhere knows that the sentence is true, which immediately contradicts the sentence and as such it becomes of no further use, yet another wrong statement among the vast oceans of other incorrect propositions. But this conundrum may be far less powerful than appears at first: the sentence does not and cannot exist in a vacuum, but it had to be conceptualized and expressed and written by *someone*. So if it is true, then it also becomes immediately wrong, because the person who expressed the sentence in the first place is a party to its knowledge. Hence, the sentence turns out to be always wrong, no matter what (35).

More embarrassingly, William Poundstone and others have convincingly shown that knowing everything at all times can in some circumstances place someone at a clear disadvantage.

Poundstone gives the example of playing chicken, a game where two people drive cars head-on at each other. If no one blinks at the last minute, the cars collide and both players die. If both cars yield and swerve sideways at the same time (it was beforehand arranged that they would swerve in opposite directions), both drivers are safe but no one wins. If only one car swerves at the last instant while the other driver keeps straight on the head-on collision course, then that gutsy (or foolhardy) driver wins. Now if one of the drivers knows everything, then *the other driver will win every time:* that other driver must simply decide to keep going no matter what, and the all-knowing other driver has no longer any choice but to always yield—or die.

Regardless of any specific scenario—it is doubtful whether a Godhead would ever want to play chicken—this example points up to the fact that some infinite traits or abilities can unexpectedly seem to lead to self-limiting features. But we are not, *by definition*, prepared to accept any curbs on an infinite Godhead without further ado or at least compelling explanations, and these apparent limitations will prove to be useful tools in further exploring the nature of infinite Godhead. In the narrow case of the game of chicken, this could be solved in a variety of ways. One way would be for the Godhead to take over the other driver's will, and hence we will first explore human *free will* below. Another way would be to maintain unpredictability as an essential, ever-present feature of reality, so that the human driver could never be sure—neither of his or her opponent's reaction, nor indeed of his or her *own* reaction, nor even for that matter of whether the message of knowledge is being relayed reliably and error-free at all times (maintaining such unpredictability as a feature of material reality is closely embedded in whether true free will can ever exist, which will be looked at below.)

There may also be other ways whereby this conundrum could be solved which we, because of the phenomenon of emergence, are unable to conceptualize or recognize. To begin with, in the realms where Godhead would dwell, there would exist no essential difference between the Godhead and the driver: the driver is a spark off the Godhead, and no one wins and no one loses. Maybe the very concept of winners and losers is a demergent feature of reality—in the wider, infinite realms of Godhead, there is no such thing, and it is a place where, paraphrasing Joel Best, everyone is always a winner. Or equivalently: to be *able* to play the game of chicken in the first place, the Godhead would have to incarnate—and thus undergo all the inevitable attendant losses, in abilities generally and in all-knowingness specifically, that stem from demergence.

This analysis can be extended to a whole category of objections that are sometimes raised in order to gainsay certain attributes of Godhead, and thence the possible existence of Godhead itself. For instance, Edward Nelson belies the possible attribute of almightiness with the simple statement that 'I can find a stone too heavy to lift, which is something that God cannot do'. To begin with, there exist seemingly legitimate questions which are in fact illegitimate—questions which seem all right and innocuous enough but which are actually quite wrong, because they are non-applicable, misleading, non sequiturs, and the like. The poster-boy for such questions is 'What came first: the chicken or the egg?' The answer, of course, is that the question, despite looking legitimate, is totally illegitimate and plain inane—because the chicken and the egg co-evolved. The question 'I can find a stone too heavy to lift, which is something that God cannot do' is equally misleading and can be answered in a variety of alternative ways.

A first way would be again to observe that in order to lift a stone, a Godhead would have to first become incarnate—we have seen that the very fact of Godhead necessarily disappears upon incarnation because incarnation entails demergence, and Edward Nelson's seeming contradiction just disappears. Asking a Godhead to lift a stone would be logically equivalent to, say, asking someone who lives in England to visit Paris without leaving the borders of England: it just makes no sense. A second way would have to go through a deeper analysis of the relationship of a Godhead with the stone—the stone is made of materiality emergent from the quantum vacuum, and, as we will be led to conclude from our analysis of omnipresence, the Godhead is necessarily present inside the stone, so that the act of lifting a stone does not mean the same thing for an actor independent of the stone (such as Edward Nelson) and for a Godhead who infuses the stone. The presence of Godhead thereby leads to a qualitative change in the meaning of the verb 'to lift' due to emergence. Our everyday vocabulary did not evolve to be able to deal with such questions, and also incorporates a measure of cognitive bias, born of our evolutionary past. It is in principle not adapted to emergent contexts, and it is this very disconnect which, for lack of a more differentiated vocabulary, allows for formulating questions which at first blush may seem legitimate but are not. This disconnect applies to most ordinary words—want, desire, intend, and so many more. If someone tells you that they know what Godhead wants, they thus happen to be wrong twice over: for starters, owing to emergence they cannot possibly know or second-guess what It wants. Second—they cannot understand either what the word 'want' truly means within the context of a Godhead.

Omnipresence

There is also mathematical proof that the Godhead, if It exists, is also inside *everything*—including inanimate things. This extraordinary result says that if a Godhead exists, it is omnipresent (sometimes called *immanent*), and hence the whole universe is in some way alive—every stone, every gust of wind, every doodad everywhere (as we shall see, this result will be found again by a quite different route when we analyze free will.)

Here's why.

If some physical attribute is to be fully known about some inanimate object, including all of its attendant characteristics and properties, there are only two ways to achieve such knowledge: either to continuously create these attributes, or to be able to see or read, or in other words to *measure* them, at all times. If a Godhead is to know all the big and crucial and determinant things in the universe or metaverse, then because of something that has sometimes been labeled the butterfly effect, It has no choice but to also have full knowledge of every single small, puny, seemingly insignificant thing or event everywhere at all times.

Should we manufacture or create some object, even to an impossibly infinite degree of precision, with well-defined and well-known initial characteristics—a number of events will inexorably and continuously alter these characteristics over time. Random local quantum events at molecular and atomic levels will continuously trigger all kinds of subtle changes. Random cosmic rays will continuously pass through the object, occasionally triggering unforeseeable tiny changes in its molecular-level make-up. Say that we manufacture a piece of string with a well-defined weight, length, elasticity, chemical make-up, tensile strength, etc.: the phenomenon of ageing, brought about by such random changes over time, will slightly but continuously alter its elasticity in not infinitely precisely knowable ways.

To fully know any object at any given instant in time, we must be able to either impose or measure all of its attributes and properties, weight, size, elasticity, full chemical make-up, etc. at that exact instant (some of these properties, such as a full precise map of its molecular make-up, would be extremely difficult for us to fully apprehend or impose, but would in principle present no problem to a Godhead. Imposing values everywhere and everywhen requires almightiness, which we will look at next.)

In terms of observing in order to know, by looking at, or in other words by continuously *measuring* all of the object's attributes, the mere fact of measuring certain attributes would make it in principle impossible, for anyone but a Godhead, to measure at the exact same time some *other* attributes: for instance, should we want to know and measure the exact length of a piece of string, we just cannot at the same time measure its tensile strength (i.e., its resistance to being pulled from both ends), because the mere fact of slightly pulling the string apart in order to measure its tensile strength unavoidably leads to an imprecision in any simultaneous measure of its length (as it slightly elongates the string.) This is a common situation in nature. Some attributes can be measured simultaneously to any level of precision (for instance the weight of a car and its color), and some others not (e.g., the car's location and its speed.) When two attributes cannot be measured simultaneously to some calculable high level of accuracy, these two attributes are called 'Heisenberg conjugate' attributes (36). The issue here is purely one of knowledge and measurement, not of existence: the speed and the location of the car, or the length and the tensile strength of the string, do exist and have well defined values at all times. It is only that we cannot possibly *measure* them, and thereby know them, at the exact same time.

Now, being all-knowing however means exactly what it says, to wit knowing everything at all times—i.e., all the values of all the attributes of everything in the universe. To be able to achieve this, a Godhead would have to bypass the restrictions and limitations imposed by the Heisenberg-type constraints described above. The only way to do so is, again, to be present *inside* the objects *at all times*, so as to be able to know directly what the values of all the attributes are, and not observe from an outside standpoint which would automatically bar full simultaneous knowledge. Because of chaotic micro-events, such knowledge is not trivial but essential, because slight changes in micro-events can soon lead to totally different macro-outcomes at human-scale situations. In turn, human-scale situations and events directly affect people and how they may react to developing circumstances, what choices they make, whether they choose the high or the low road, and so on. A hurricane triggered by some butterfly somewhere has the power to test the locals' mettle and fortitude in more ways than placid weather would, and to change lives. The impact that slight random events can bring about in real life is embodied in the stock phrase 'life turns on a dime', and has furnished a recurrent theme in art and literature. It is even the theme of the only ancient Roman-era novel that has survived in its entirety to the modern day (Apuleuis's '*Metamorphoses*), as well as that of innumerable movies (we may think here of Woody Allen's '*Match Point* and many others), and novels.

To qualify as a Godhead, there is therefore no way around knowing everything at all times: should It not be present everywhere, there would immediately exist huge numbers of attributes, characteristics, properties of the physical universe that the Godhead would soon know nothing about, and this ignorance with its trailing wake of material consequences would quickly balloon out of control.

A possible objection would be that a Godhead would not necessarily need to know everything at all times, and could simply choose to somehow go back in time to examine something whenever It would, for whatever reason, need to know something It would have happened to overlook. It does not work, on a variety of grounds. First, because of chaotic effects the complexity and baggage involved in going back in time would be far heavier and unwieldier than is the case in simply being present everywhere at all times. Second, such an MO would depend on a number of assumptions about spacetime itself. If we agree with Einstein's view of the equivalency of space and time, then we are back where we started, because we'd be saying that the Godhead must be present every*when*—at all instants of time—and there is then no material or qualitative difference in saying that the Godhead is present everywhen rather than everywhere (37). Should we take however a non-Einsteinian view of time, then a whole cornucopia of even more complex scenarios arise: to go back in time, the Godhead would then variously become faced with a vast or infinite number of possible pasts to go back to, and the ability to know which particular past to go back to would be the logical equivalent of already knowing the particular past in full: in other words, in this scenario, our attempt to dispense with the burdensome need to know fully the present at all times would lead to the far more burdensome necessity of fully knowing several pasts. Whichever way we turn, we cannot bypass, as a necessary feature of Godhead, the requirement to be present everywhere at all times.

Beyond being everywhere within our universe, where else is the Godhead present? The picture so far is that of a Godhead who has infinite traits, such as an infinite IQ, lives in many dimensions, and is not material in the ordinary sense of physical materiality (38). Its presence ensures that the universe, or the metaverse we ultimately live in, must be infinite, because it accommodates an element, Godhead, who Itself is infinite.

This requirement for a wider multiverse is borne out by a totally different route, as follows. The maximum number of elementary actions—of separate elementary *doings*—that can ever be carried out in our known universe is easily calculable. By 'elementary' action is meant any smallest 'non-nil' occurrence, any smallest conceivable micro-event, such as an elementary particle moving from its current position to a next available location nearby. The calculation takes into account the overall grand span of time available during which elementary events can happen within our universe—an expanse of time that goes from the birth of the universe to its ending (i.e., some future time when either all matter has become hopelessly diluted and inert in a 'Big Freeze' type scenario of the end of the Universe, or when the Universe ends in some other way.) The calculation also presumes the optimized use of the fastest possible speed of interaction between any participating material items, such as particles or atoms, the shortest possible distances for some micro-event to occur, the smallest possible material constituents, and so on, capable of leading to some elementary action or event taking place. With all these ideal conditions in place, the calculation shows that the maximum number of elementary micro-events ever available within our finite universe is about 10 to the power 168—10 followed by 168 zeros (39). In turn of course, any larger-scale action or chain of events or happening is a composite event made up of a finite chain of a number of elementary actions—much like walking a mile results from taking a number of individual steps, and in turn each individual step is the result of a large number of more elementary actions, such as neurons firing in the brain, muscle fibers being tensed, and so on.

The overall number of possible events in our universe is therefore far, far bigger than the mere number of all possible micro-events: any event made up of constituent separate micro-events, taken from the pool of the available 10 to the power 168 elementary events, constitutes a new separate event unto itself. (Separate such compound or macro-events can also overlap, i.e. freely make use of elementary events also involved in other macro-events.) Whereas the overall number of possible macro-events is incalculably large, it is however *finite*, a direct consequence of mathematics (specifically of enumerative combinatorics) applied to the finite set of micro-events of which all macro-events are ultimately a combination. (We cannot even begin to speculate on that number. It has been noted elsewhere that the mere likelihood of, for instance, any given novel being actually ever written involves odds in the range of one over ten to the power googols of googols of googols .... an incalculably gigantic number.)

Despite their mind-boggling hugeness, all these numbers are actually *vanishingly small* in comparison with infinity itself, at any level of cardinality. Combinatory analysis thus leads to an insight that it is *impossible* to build infinity starting from a finite basis or environment. From a pool of any finite number of elements, we can only ever combine them in myriad ways whose numbers, however vast, can never breach into full-fledged infinity.

However we may try, the ability to create infinity from finite circumstances is firmly barred. The reality of mathematics and of enumerative combinatorics always imposes itself, however circuitously. For instance, one way we could try to to make infinity real—to 'import' unrealized infinity into actual reality, as it were—would be to focus thoughts specifically (however fleetingly), on *every single* number within an endless series. But owing to the limitations of materiality and of finite time, we cannot do so. The only way we could succeed in this scenario in importing infinity into reality would be if infinity already existed—in the form of an infinite amount of time. In simple terms, this means that the nature of our existing reality is such that *either* actual infinity exists, *or* it does not, and this status of either existence or nonexistence can never be changed. Because of straightforward combinatorics, if all we have is *potential* infinity (such as the endless series of all the natural numbers), as opposed to *actualized* infinity, we will never be able to parlay that unrealized infinity into existing reality.

Should we however start from a pre-existing base of already available infinity, these same limitations will also curb our ability, if not a Godhead's, to *exercise* that infinity. For example, there exists an aleph-2 infinity of possible geometrical curves, yet we will never be able to build an infinity of them, irrespective of how fast and how long we work. The only way to do so would be to either to generate such curves within zero time, *or* alternatively we would have to keep generating such curves over an infinite expanse of infinite time so as to achieve an infinity of them, and we'd then be straight back in a situation where infinity itself pre-exists (40).

There is thus a stark, impregnable wall between the infinite and the non-infinite: infinity can never be created out of whole cloth from within any finite reality, and therefore to ever exist at all it must pre-exist to begin with. The *theoretical* question of whether infinity actually exists in outside reality thus becomes circular: it exists if it does, and does not if it does not. This has a direct bearing on any *theoretical* answer to the question of whether a Godhead exists or not, debated since time immemorial but never compellingly answered either way: the question is *unanswerable* by logic alone. Infinite Godhead exists if It exists, and does not if it does not. If anyone ever tells you that 'science proves that Godhead does not exist', or alternatively that 'science proves that Godhead exists', it is simply wrong. In either case, it cannot be true *in principle*. Godhead, the ultimate possible case of emergence, simply cannot be apprehended by logic from within a finite world. Many people have been led to the same conclusion from a variety of other angles, not only mathematical (Georg Cantor), but also philosophical (Richard M. Gale, et al.) We are led straight back, through a roundabout route, to Cantor's insight that any possible existing Godhead can never be perceived by logic alone, and that our only hope to ever possibly 'put out our hand and approach the face of Infinity' must be through intuitive insight.

This begs the question of whether Godhead could be apprehended by *experiential* or empirical means—either directly or indirectly. Since we cannot *in principle* directly observe or measure infinity with any measuring apparatus from within a finite environment, because infinity cannot be properly mapped onto the finite, *direct* observation is ruled out. Even if we could somehow apprehend a glimpse of infinity from our finite base, this glimpse could only be *infinitely* smaller than the real thing, infinity itself. Therefore, unless we can compellingly and objectively ascertain Its existence by indirect observation, the *Godhead Itself becomes a non-observable*, in the scientific meaning of the word. (We will be looking at the possibility of indirect observation.)

- A non-observable is anything that cannot, due to its very essence, be observed, grasped or otherwise apprehended fully independently of itself, because observing it always ends up involving the phenomenon itself. In physics, time is an example of a non-observable. Time cannot be measured independently of itself, because all our attempts to do so, for instance through the sweeping hand of a clock, end up directly or indirectly involving either space or time or a combination of both (such as in some phenomenon's speed) which are components of spacetime and hence of time itself.

The existential wall between the realms of finiteness and infinity also demonstrates that if a Godhead exists, *It can never become entirely incarnate into a non-infinite environment.* To try and do so, It would have to somehow squeeze infinity down into finiteness. (If, against impossibility, It somehow still managed to do so, it would have to cross the wall, which would result in irrevocably relinquishing Its infinity. It would thereby become unable to ever return from finiteness to Infinity, which would by then have simply ceased to exist.) This constitutes type-A proof that if a Godhead exists, It can only ever interface with the finite world indirectly, perhaps through manipulating matter, or perhaps through soothsayers and other sundry 'sons of God'. Combined with the earlier proof of the inevitability of different renditions, this shows that *even if a* or *the Godhead exists*, clashing interpretations of divinity are inescapable in any finite world—they're built-in in the way Infinity becomes distorted and garbled when seen through the lenses of a finite world. We have also seem to have found something—incarnating fully—that a Godhead actually *cannot* ever do, because doing so would end Its very Godhead (41).

To be able to exercise Its infinity, a limitless Godhead must by definition be at liberty to perform an infinity of actions. *To exercise that freedom, It therefore needs at least either an infinite universe or an infinity of finite universes.* A Godhead stranded in a lone finite Universe or in some limited multiverse would become curtailed, relegated at an infinite remove from Its own potential infinity—for ever unable to exercise full infinity, it would therefore *not* be infinite by any workable or acceptable meaning of the word, and hence would be less, by any definition, than a full Godhead. There is no reason why a Godhead should countenance such drastic curtailment of Its Godhood, and indeed doing so would detract from Its very divinity: some form of infinite universe or multiverse is the only reasonable nature of reality if there is a Godhead.

A question that naturally arises is that of how can a Godhead be present inside, say, a murder weapon? Before we address—and resolve—this question, we must examine one consequence of omnipresence, to wit, how it leads straight to the quality of almightiness. In turn, as we shall see, almightiness feeds back in an unexpected way on the quality of all-knowingness.

Almightiness

The question arises whether a Godhead present everywhere and every*when* would remain a 'passive' observer—i.e., whether It would somehow *see* everything but let things take their own course without necessarily subjecting them to intervention, whether in whole or in part—or whether It would determine everything, every event, every outcome, every development and every unfolding of events in time.

There is also the question of how something immaterial, such as a non-physical Godhead, can affect materiality.

To find answers, we must call upon quantum mechanics, the most successfully predictive and explanatory scientific theory to date. It shows, first, that materiality itself turns out to be an emergent phenomenon squarely born of immateriality—of the immaterial quantum vacuum, also known as the quantum froth, which pervades everything (*see note* 7). Second, it hints at a still controversial and not fully understood possible role of consciousness in weaving the fabric of reality—an effect known in physics as the 'observer' effect. The primary role and material influence of an observing consciousness is present in a number of interpretations of reality (at its core, the 'observer' effect simply arises from the act of measurement where the observer takes on the role of the 'measuring system', *see note* 36.)

The role of some consciousness is sidestepped in other interpretations, such as the Everett many-worlds interpretation of quantum mechanics. One issue with most of these interpretations is that they only manage to avoid a role of consciousness in the co-weaving of reality at the exorbitant cost of very unwieldy baggage, such as full-fledged, ready-to-go parallel universes popping up continuously into reality like hot pop-corn. Even if reality works that way, it does not help much, because the largest (infinite) number of universes which could ever be created that way would be of the relatively low cardinality of aleph-1. But Godhead is, we have decided, of aleph-infinity cardinality, infinitely stronger than aleph-1. Therefore, even Everett's endlessly geysering universes would be too small to accommodate a Godhead: if It exists, a godlike wider multiverse must be infinitely bigger than Everett's multiverse. Everett's scenario of branching pop-up universes constitutes physics' best attempt to escape the inherent indeterminism and unpredictability built into quantum mechanics. Yet the above shows that, even if it were a correct picture of reality, *and* if there is such a thing as a Godhead, then Everett's multiplying universes cannot be the whole tale. (We shall revisit this theme when we further analyze free will, the openness of infinities, and the mathematics that opens the door to the possible existence of evil.)

Beside the Einsteinian view of space and time or Everett's view of a continuously multi-branching universe, there also exist other interpretations of reality, encompassing some thirty-odd interpretations of quantum mechanics, each fraught with a specific take on reality and on how the universe or metaverse works (the recent 'objective collapse theory' perhaps constitutes a more promising avenue of exploration; it eliminates the observer effect to replace it with a naturally occurring statistical effect.) What all these possible views of reality agree with, however, is that the *measurement* of any attribute of anything has an immediate material impact on the value of that attribute. As we have seen, any *observation* is always, technically, a form of measurement, and in turn things and events are always materially affected by observation.

Through adjustable or modulated participatory observation, observed events can thus be steered at will (which may include the option of letting them evolve on their own.) But this is the very definition of almightiness: all-knowingness has inevitably led to omnipresence which in turn has led to the immediate ability to steer events at will.

The qualities of omnipresence, almightiness and all-knowingness thus indivisibly hang together: if any one of these is present, the other two are as well. Where does that leave the possibility of free will for mere mortals if there is such a thing as a Godhead?

The Free Will Theorem

There exist many studies of human free will from different angles, and even a cursory search will turn up scores of books and monographs dealing with the subject. Yet, the scientists who have investigated the subject in-depth cannot agree on whether it exists, in full, in part only, or not at all. Unsurprisingly, there are wide shadings of opinion in the camp of those who believe that we only enjoy a limited measure of free will. For instance, there is compatibilism—the slightly odd notion that determinism and free will are not incompatible, that we live in a fully foreordained universe and yet are still somehow able to exercise a measure of free will. Different nuances of compatibilism exist, which mostly differ in how deep determinism is thought to reach. The most common view of compatibilism is reflected in Schopenhauer's phrase 'Man can do what he wills but cannot will what he wills'. In some renditions, it becomes similar to libertarianism, which favors a measure of free will but under whose view the range of choices available at any juncture is narrowly determined by factors beyond the subject's control (we are reminded here of best-selling author Stephen King, who answered a student's question as to why he didn't write in other genres than horror with "What makes you believe that I have a choice?") Examples that can be viewed as supporting libertarianism abound— Beethoven only ever composed *Beethoven-like* music, for instance, although within the demonstrably infinite (but bounded) ambit of his music, he was probably free to compose anything.) The bottom line being that no one, despite strong opinions on all sides, can make a compelling, 'smoking gun' case either way. What is nevertheless established is that Heisenberg unpredictability, deeply embedded within physical reality, opens the door for the physical possibility *in principle* of free will (42).

Some have argued against the possible existence of free will as follows: free will can only play out either through quantum interactions (which are nonpredictable but random) or macroscopic interactions (which are predictable but non-random). However by definition free will would have to be both non-random *and* nonpredictable: hence, free will is illusory. This argument seems open and shut, but it is wrong. Quantum effects never cease to operate, including at macroscopic scales, although they may be hidden from plain view by statistical effects (indeed, so many particles are involved in any macroscopic object or event that overall, outcomes gravitate towards wholly predictable statistical values.) Indeed, most of today's tech gizmos (such as transistors, computers, lasers, DVD players, and so on), are squarely based on purely quantum effects, and they work in macroscopic-scale environments.

Others, such as the cognitive scientists Michael S. Gazzaniga, Daniel Dennett, Daniel Wegner, Victor Lamme, et al. use other arguments to deny outright the existence of any measure of free will. The kernel of their arguments ultimately rests on an original experiment by Benjamin Libet, which showed convincingly and repeatedly that a supposedly free decision made by subjects between several available courses of action, was actually made within their respective brains (as measured by brain MRI scans) *before* the instant the subjects said they had freely made up their mind—a full third of a second or more before. However, it can reasonably be argued that their interpretation of this experimental result probably rests on the oversight, perhaps, of other relevant data, and even on a number of unrecognized implicit assumptions. A recent paper by Jesse Bengson at the University of California at Davis argues that brain noise, inevitably arising from the brain's electrical wiring, opens up random micro-events which lead to 'cracks' in predictability that allow for free will to insert itself in. As Bengson puts it: "Brain noise inserts a random effect that allows us to be freed from simple cause and effect." Interestingly, Bengson's work, far from gainsaying Benjamin Libet's experiment, *builds* on it and takes it several steps further: instead of relying on the experiment volunteers telling when they made up their minds, Bengson's experiment was designed to identify objectively when conscious decisions are made.

Another major issue in the interpretation of Libet's results as supporting the non-existence of free will is its implicit assumption that time flows smoothly from past to future. This rather common assumption is born of everyday experience, but it leads here to a possible case of unconscious but, as it turns out, unwarranted cognitive bias. Indeed, well-known repeatable experiments show that time simply does not linearly proceed from past to future. A class of experiments known as *delayed choice experiments* shows that inert nature itself can, quite surprisingly, reliably opt for a given course of action *before* the totally random decision is made which will set the given course of action in question. Something similar could occur in the brain, and the firing of the neurons in the brain's decision circuits which lead to a particular decision being taken could also take place within some complex, non-linear time loop, in the same way seen in delayed choice experiments.

Other scientists, such as Mark Balaguer, Robert Kane, Evan Harris Walker, et al., have argued in favor of free will.

- From a narrow mathematical perspective, if we take the reasonable view of wave function (43) realism, we would agree with them: because there is no possible way whereby a picture of the mind as being governed by wave functions, and hence subject to the gaps in determinism brought about by the Heisenberg picture of reality, could be reconciled with full robotic predetermination. Moreover, anything permissible tends to happen one way or another in reality: as the saying goes, 'nature abhors a vacuum'. Since physical law leaves the door open to the possibility of free will, the balance of evidence seems to favor its actual existence.

Many others still, such as neuroscientist David Eagleman, just keep an open mind. And then there are also those who think that the debate about free will is the wrong discussion to begin with, a debate which deals with the form, as they see it, rather than the underlying substance (Thomas Metzinger, et al.)

An intriguing side issue in the discussion on Free Will is, somewhat unexpectedly, the role of parasites. Parasites have been shown to partly or in some cases entirely hijack the will of much higher life forms for their own ends, which hints at Free Will being perhaps so totally predicated on biochemical processes in the brain that it may ultimately be elusive. The issue has been looked at in some depth (Kathleen McAuliffe, 2016, Carl Zimmer, 2001, et al.) but no definite conclusion could as yet be reached.

For our purposes however, the most telling argument is something called the Free Will theorem.

In simple terms, the theorem demonstrates that under certain conditions, *if* we do have free will, *then* individual elementary particles must *necessarily* have a measure of unpredictable free will as well. The theorem's argument is a simple one to follow, and it is compelling. In essence, it says that if an observer is truly free to make a decision of some kind, then, provided certain conditions are met, *this decision cannot possibly have been determined or foreordained by anything that has taken place anywhere in the universe at any time before the instant the decision was made.* Ultimately, it means that if free will exists, no physical object, no particle anywhere in the universe, could possibly carry within itself some feature or attribute that would have ineluctably led to the particular decision made. The consequence of this statement can be drilled down all the way down to the level of a single particle: the possibility of free will ineluctably leads to the capability, on the part of matter itself and therefrom the simplest constituent of matter, to exercise a measure of free will. In other words, the theorem says that if free will does exist at the level of the observer—i.e. the lab technician, or you or me—then it also inevitably exists at the elementary particle level. This does not prove free will, but it compellingly demonstrates that free will exists, it trickles all the way down to the level of the elementary particles of matter, for the obvious reason that since, by definition, free will and its exercise must remain free of anything that has gone prior to the instant of its exercise, then matter itself has then no choice but to be complicit in allowing this to happen. Since a Godhead would by definition have free will, this result can be seen as fully compatible with the earlier demonstration of a Godhead being present everywhere, and with the notion that if there is a Godhead present everywhere, then there exists a mechanism whereby free will has trickled down everywhere and imbued everything.

A Conclusion to Part 1

The picture that emerges so far is that of a more complicated essence of Godhood than the pictures most creeds and clerics have traditionally presented us with. The most basic tenets of faith which virtually all monotheistic cults and religions everywhere have always asserted, seem corroborated by simple mathematical analysis based on the prior assumption of a Godhead. Assertions such as 'God is almighty', 'God sees you everywhere', and such, are not trivial but lead to mandatory consequences, some of which may well stand in ultimate contradiction with *other* statements also asserted by the same cults. Before we go on to probe deeper into the nature of Godhood, and, for instance, understand why a Godhead could be present in a murder weapon, we must first look at a few widespread dogmas.

PART 2

He Said, She Said

As long as the world's great spiritual traditions confine themselves to describing a few basic inherent *attributes* of any possible Godhead, they seem, from a mathematical angle, to be largely right: *if* there is an infinite Godhead, then yes, the Godhead is everywhere, all-knowing and all powerful. From a foundational assumption that Godhead exists, then these assertions are also borne out by the math (44).

But the math also compellingly shows that we can only hope to understand the essence of an infinite Godhead by developing empathy for Its inward, *immanent* qualities. We cannot understand Its *outward* actions, nor how It would *think* (for lack of a better, more multi-dimensional word), or what it would do or want to do. We are inherently totally unable to second-guess It. To paraphrase a former US official, there are the things we know, the things we know that we don't know, and the things we don't know that we do not know—and above and beyond that there are the things that even if we were shown and told and tried, hard as we can, to grasp, we still *couldn't* possibly come close to 'getting' them, because all the limitations inherent to our physical state would prevent us from doing so. The far-and-away immense majority of what a Godhead would know or would be able to do, firmly belongs in the latter two categories, including the items we are not equipped to even begin to imagine let alone comprehend. And when men somewhat overweeningly attempt to do so, the results can be random, sometimes horrid.

For instance, we established that, mathematically, any possible Godhead must be present within us—and indeed that if It weren't, it would conclusively and mortally detract from Its very Godhood. Yet, the belief in oneness with the Godhead is still deemed false ('heretical') by most established monotheist creeds, and historically at least, such a belief was apt to be harshly sanctioned. As a case in point, because the ninth-century Sufi mystic Al Hallaj wrote "*I saw my Lord with the eye of the heart, and he asked, 'Who are You?' I replied, 'You*", a wonderfully poetic way of saying the very same thing that the simple earlier mathematical proof established, he was put to death. The emergence and sharp discontinuities that crop up when Infinity appears onto the scene were utterly alien to the finite human minds of Al Hallaj's judges and executioners. To their way of thinking, Al Hallaj's words betokened a willingness to usurp an undue spot at the top of some worldly hierarchy, much like a commoner pretending to be king would be guilty of lèse-majesté. For this perceived sin, he had to be put to death.

Clerics throughout History have believed they could know the mind of God, and wading through the morasses of sundry dogmas, rules, edicts that have been promulgated over the course of known history can on occasion make for a singularly depressing experience. Some exhibits are harrowing: here are the Aztecs, blithely slaughtering tens of thousands yearly lest, according to their priests, the sun petulantly refuse to rise in the morning, in a childish pique of high dudgeon at the absence of bloodshed. Over there in the corner, the Inquisition and its trailing wake of zealous, unspeakable torturers-with-gusto henchmen raging across Europe through long periods of the Middle Ages. In yet another corner, fiery Taiping true believers hunting down and murdering in inventively gruesome ways tens of thousands of men, women and children in Nanjing in 1853. The list goes on and on. With so many different dogmas and interpretations and shadings of dogmas out there, we will routinely find, should we care to look closely enough, peremptory assertions and their very opposites being promulgated by sundry religious cults at different places and times: for instance 'making and/or listening to sacred music is good/a duty', *vs.* 'music is a frivolous abomination in all its forms and must never be partaken of', or 'sex is (*sometimes/always*) bad and abasing' and 'sex is (*sometimes/always*) good and uplifting', or 'one should learn as much as possible of the world to better understand the works of God' *vs.* 'one should not learn anything, because learning anything detracts from the duty of exclusively focusing one's mind on God'.

Sometimes the dogmas are freely interpreted by the priestly classes, not necessarily in keeping with the written canons of their respective organized faiths, and in some cases plain made up. Yet, it is the very existence of the cult that opens up the possibility of arbitrary or untoward interpretations. While researching this book I interviewed a number of clerics from a range of different creeds. One of my questions was 'what do you think of people who, in their lives, practice S&M (sado-masochism)?' Down to one notable, single exception, all the others spoke scornfully of SM practitioners in uniformly judgmental, scathing, and sometimes surprisingly gleefully damning terms: "SM practitioners are lost souls, sinners, they should grovel before God and ask for forgiveness, they are victims of the devil, they shall be thrown into the lake of fire..." You get the picture.

Yet, unbeknownst to the clerics, my question was disingenuous: science has something very specific to say about SM practitioners. Robert Stoller, a California doctor who investigated a statistically significant number of people practicing hardcore sado-masochism (45), discovered that *all of them* bar none had happened to experience excruciating medical issues in their early childhoods—issues that had led to a number of surgical interventions and lengthy hospital stays. The little children they were then had been so physically hurt and terrified that, in order to be able to simply stay alive, their brains had rewired: to trick the child into surviving, despite the terrifying impersonal grown-up doctors in white coats, despite the operations, the blood and the pain, the brain had rewired itself to associate the pain and terror with ..... *pleasure.* To save the child, the brain outright lied: *it's not pain, dear: it's pleasure.* That did the trick: the children survived. A long-term consequence, however, arose because the brain is just so much more plastic and malleable during its formative early years than later on. By adulthood, the association of pain with pleasure had set, and henceforth become a constituent element of psychological health: the brain jelled into adulthood with a now hardwired need for occasional pain so that psychological balance be maintained.

And there we have it: science's deeply explanatory and humane approach towards people who can certainly use understanding, compassion and respect for what they went through and most often just barely survived, and a peremptory, shallow and most definitely ungodlike approach by many clerics. The difference could hardly be starker.

Throughout History, men with perhaps a worldly agenda, or perhaps simply way too much confidence in their own counsel, have co-opted the people's yearnings for transcendence and ensnared them in a tight made-up mesh of sometimes nonsensical rules, dictates, obligations, often enforced by the all-too evident threat of eternal damnation or even outright worldly violence should unquestioning conformism not be sufficiently forthcoming. But the straightforward bottom line reality is that any Godhead is either infinite and hence inherently not possibly into any form of pettiness, or It is *not* infinite, and then probably not a Godhead. As we shall see in Part 3, if there in fact is a Godhead, then the only edicts, ukases and dogmas that can ever make sense are those that happen to be mathematically conducive to the demonstrably indispensable growth of a specific form of infinity. Those which lessen the overall volume of hurt in the universe are the most contributive, because as we shall see hurt in general seems to lead to stunted growth rather than to expansion.

Even a cursory list of the arbitrary dogmas issued by priests from various and sundry cults who, throughout time and the world over, have proclaimed that they know better and assumed self-bestowed divine authority and the right to tell others how to lead their lives, would make for singularly glum reading. Kevin Dutton, a psychologist who spent his career studying psychopaths, found that the desire to dominate people, and dictate how they should live their lives, correlates highly with psychopathy. Correspondingly, the professions with the least number of psychopaths are whose job consists in delivering non-judgmental help and assistance to others: care aides, nurses, therapists, beauticians, charity workers, teachers, and medical doctors all make the grade into the top ten occupations with the fewest psychopaths. On the other hand, clerics happen to score within the top ten occupations where the highest number of psychopaths are found.

We will only select and analyze logically a few, in a bid to keep this painful chapter as short as necessary. We will pick specific dogmas either because they would appear to be the least obviously inane or may even look reasonable, or alternatively because a modicum of analysis will unexpectedly show that they may hide a precise although unapparent agenda.

Sometimes mitigating scientific explanations to imposed dogmas have been sought, and sometimes these explanations do make sense (a case in point would be the ban on eating foods prone to unhealthful contaminations, presented as a religious obligation, or, say, a ban on using the same hand used to wash oneself to then eat, especially in areas where water tends to be scarce), and sometimes they do not hold up to scrutiny (e.g., an explanation was attempted to explain large-scale Aztec human sacrifice and follow-on ritualized cannibalism by an unrecognized, latent need for animal protein, an explanation which however did not stand up to analysis.) But even when hindering people from doing something may be, for their own good, justified, would not the truth itself, rather than couching the ban in some made-up tale, be far more dignified?

A frequent explanation for dogmas, and an inescapable contributive part of the reason why they so consistently and consistently arose, consists in the historically frequently present need on the part of the ruling classes to keep a measure of control over the populace—if need be by stoking fear, enforcing ignorance, and backing up their claims to power by invoking an authority far beyond that which would otherwise accrue from their ordinary humanness. Sometimes, as Kevin Dutton suggested, the desire to control other people's lives may be wanton, rather than merely utilitarian (when seen from the vantage point of the ruling classes.)

Historically, those instances of transcendence which ordinary people may have experienced from time to time, such as the mystical experiences that various people have reported throughout history, may have led to a natural need for a measure of organized spirituality on the part of people seeking to make communal sense of their experiences, all too often readily co-opted and channeled into organized grooves by earthly powers. The deliberate but false amalgamation between whatever true instances of spirituality may exist on the one hand, and formally organized rites and cults led by the ruling classes on the other hand, has been forcefully encouraged and nurtured throughout history, although it is logically inescapable that the two have very little in common. This deliberately induced confusion of the legitimate spiritual needs and longings of people with the elite's power agendas has been foisted upon societies since time immemorial, and often abides to this day.

Indeed, if we look back at Europe at a time when official organized faiths had a much bigger role and influence in society than they do today, we find that the strictures they imposed on certain private domain behaviors, such as sexual mores or eating habits, were consistently shrugged off and willfully ignored by the self-appointed guardians of the faith, most often the upper classes and royalty (46).

A number of studies indicate that, irrespective of era or location, up to about a half the population anywhere and anywhen have experienced *something* which they felt had a deeply spiritual or transcendental origin or meaning at least once in their lives. When faced with the occasional hard facts of life, such as the death of a loved person, most people will feel the need for some marker—some ceremony, a solemn moment of stillness, something. Illustrating these statistics, any 'open line' type radio shows or magazine article featuring stories of transcendence invariably draw out many people who volunteer tales of transcendence which they had hitherto kept for themselves for fear of ridicule, or of being branded a liar or a weirdo. The industry that caters to spiritual longings in some shape or form is, in the Western world, a multi-billion dollar industry. The temptation for the powers that be, or the powers that want to be, to co-opt and use to their own ends such widespread spiritual yearnings and/or experiences can prove irresistible, and in our modern world, the forcible amalgamation of spiritual longings with the secular interests of worldly powers continues to happen without fail, including in Western-style democracies. When contemporary author Sam Harris wrote at length in 'The End of Faith' (2005) on what he perceives as the fancifulness and mendacity of most established churches, he was promptly tarred as being deeply 'unspiritual' and dubbed a member of a 'unholy trinity' (with Richard Dawkins and Christopher Hitchens), even though he had specifically written: "*Spiritual experiences are real and significant.... we must not renounce all forms of spirituality or mysticism to be on good terms with reason*", and "*The truth is that we simply do not know what happens after death. The idea that brains produce consciousness is little more than an article of faith among scientists, and there are many reasons to believe that the methods of science are insufficient to prove or disprove it.*" He also wrote, after citing verbatim some of the more bizarre official dogmas of a well-known church: "Is there any doubt that, if he were the lone one, a subscriber to such beliefs would be considered as *mad*? Rather, is there any doubt that he would *be* mad?" From the standpoint of even the simplest mathematical analysis of the cited dogmas, it is very hard to fully disagree with him. For daring to express such views, pertaining solely to the organized dogmas of a church rather than to spirituality *per se*, he was immediately smeared as deeply unspiritual—because he was perceived as attacking the worldly interests of those who use spirituality as a tool to advance and preserve their secular ends.

This amalgamation is reinforced by a tacit de facto agreement between somewhat strange bedfellows—the people at both ends of the debate, those on the one hand whose stock-in-trade consists in exploiting and harnessing spirituality and intimations of Godhood to their own worldly ends, on the other hand those who deny all spirituality and seek to enforce a strictly reductionist and materialistic view of life, all too happy to point to the ridicule of some aspects of organized religion as a means to deny spirituality itself. Thus, the amalgamation of established religions with spirituality endures in society's narratives, for it lies in everyone's interests. Yet it is of course, logically and mathematically, thoroughly untenable.

In this debate, further intellectual dishonesty is rife, and glaring examples can be found on both sides of the debate. There is for instance the militant atheist Richard Dawkins, blaming the persecution of the scientist Alan Turing for homosexuality at the hands of the then British government on religious belief (in his 2006 book 'The God Delusion'.) But in actual fact it was the psychiatrists of the time, laboring under the sway of the extremely materialistic (and deeply flawed) science of the 1950s, who were to blame for that persecution—which had nothing to do with religious thought. Likewise, several cases of apparent straight dishonesty on the part of famous *über*-skeptic and professional debunker James Randi have been documented in various accounts, including accounts penned by strong skeptics (47).

On the other side of the aisle, many practitioners—vendors—of spirituality, including some well-known ones, are prone to affirming the weirdest things just on their say-so, without analysis, justification, or corroboration whatsoever. They seem hardly better than outright scammers—as in, "we'll tell you what you so obviously want to hear, and you'll give us your hard-earned money for the privilege." To borrow a phrase from John McEnroe, they can't be serious.

The London-based International Humanist and Ethical Union (IHEU), a non-profit observer of freedom of thought across the world, publishes its 'Freedom of Thought Report' on the state of religious freedom across the world every year on December 10th (timed to coincide with the International Human Rights Day). Its December 2014 Report indicated that a number of governments across the world stepped up hate and repression campaigns against non-observers of state-approved faiths. Amongst other worrisome developments, the report states that 'the overwhelming majority of countries fail to respect the rights of free-thinkers', adding that 13 states have made apostasy (leaving or changing one's religion) a capital offense. The report also indicates how 2014 was marked by a steep rise in the number of officials and political leaders agitating against non-religious people 'in terms normally associated with hate speech'. A number of countries specifically named non-religiosity (unsurprisingly deliberately, albeit wrongly, equated to atheism) as threats to society and/or to the state. Even in Western democracies, such as in the UK, specific events such as certain government-led initiatives in education, show that freedom of thought as relates to religion has gone backwards. Even in unexpected places such as Russia, public expressions of disbelieving views can be equated with blasphemy under the law and criminalized (48). In 2016, the Russian blogger Viktor Krasnov was put on trial and faces up to a year in prison for denying the existence of God in an argument on social media.

At the other end of the attitudes spectrum, official science seems to be, oddly enough, every bit as inflexibly set in rigid see-nothing, hear-nothing ways. The amalgamation of spirituality with formally organized faiths and their trappings (often rigidly formatted rites which may involve rote recitals, singing, garish ceremonials and mandatory official costumes featuring sometimes fanciful robes, hats, and sashes, and the like), and the fancifully magical claims of some of their tenets, often seemingly bereft of any discernible spiritual content, has much contributed to turn any involvement in anything that even remotely smacks of religiosity, and therefrom of spirituality itself, into a merciless and immediate career killer in science. Compounding the wreckage, the insistence of a few ignorant clerics on invoking a Godhead to glibly explain everything under the sun, and worse still as an excuse to foster a culture inimical to logical thinking and scientific inquiry, to asking questions and striving for progress in understanding, has further obviated any possible mutual respect and peaceful co-existence that could have sprung up between spirituality and science. Modern day scientists will therefore as a rule not touch anything that smacks of spirituality with a million-foot pole—never mind that most scientists today do *not* espouse a materialist, Aristotelian view of reality, if only because materiality itself ultimately emerges from unfathomable quantum nothingness (*see note* 7). It is a regrettable state of affairs, if only because spirituality, and its hijacking by worldly interests and unwarranted channeling into predefined societal grooves, has been such a key driver of many of the defining events of history, and *continues to be so*. It has shaped the organization of societies, and as such, it is in urgent need of proper, balanced factual study, and possible reshaping.

In this context it is telling that a number of world religions demand a blind 'leap of faith' on the part of their followers—and decree that questioning tenets, and the application of intelligence and of personal reasoning in the seeking of truth, are deadly sins (49). Finally destroying in the eyes of scientists the last shreds of respectability that many cults might have hung on to, the above-cited long insistence of many creeds to explain features of the world and the material universe through the agency of some Godhead abides to this day. This rather naive—and ignorant—approach to reality has become known as 'the God of the Gaps'. It has, of course, a long history—every time mankind did not understand some natural phenomenon, an earthquake, say, or a bolt of lightning, some Godhead was conjured up to explain the phenomenon (50). In ancient Greece, Zeus was both the cause and the wielder of thunder and lightning. Then, science developed and gained ground. It worked out the straightforward electrical mechanism of lightning, shut another gap in our knowledge, and the God of the Gaps retreated yet again, into ever-shrinking pools of leftover gaps.

- To the discharge of early mankind, Psychologist Deborah Kelemen correctly notes that ours is a complex environment, where the understanding of even common natural phenomena (such as thunder, bolts of lightning, earthquakes, the sun, etc.) requires a level of sophistication way beyond our natural and original abilities to understand. Hence, in the absence of the sophisticated knowledge that can only come from a relatively advanced civilization, attributing such phenomena to some Godheads is a natural, and even a sound, coping mechanism. Those explanations developed because they were instant and immediately usable—they did not require the lengthy and sophisticated educational buildup that, for instance, being able to explain lightning through the phenomenon of electrical discharge requires. Ara Norenzayan, Nancy Ellen Abrams, and many other sociologists, anthropologists and observers of religion conclude that traditional cults have played a very useful, if often mendacious role, but that of course, that role is now totally outlived.

Science kept advancing, more explanations were found, gaps kept shutting. One of these gaps was the very inception of our Universe, the Big Bang—seen by some clerics as the ultimate refuge of Godhead, a last prize bastion of unassailability. Then modern physics went on and explained not only one, but several workable mechanisms whereby Big Bangs can occur (to the extent that virtually no one anymore believes that the one Big Bang we know of is in any way unique or remarkable.) When Roger Penrose, Edward Tryon and others submitted credible scenarios for how a Big Bang could occur and how a universe could be born out of nothingness, all that was left were a few gaps still sprinkled here and there in the landscape of physics—gaps that few doubt shall slam shut some time soon. As Stephen Hawking so aptly put it, "Science doesn't prove that there is no God, only that God is not necessary."

Across different faiths, most promulgated edicts and dogmas classify into a few recurring categories, which seek to impose behavioral patterns in key areas such as clothing, food, and sex, and enforce limits to knowledge. On the positive side, some dogmas have historically managed to enforce standards of health, hygiene, and even morals, sometimes however at the steep collateral damage price of severely curbing education and, perhaps, predisposing some members of the public to unquestioning gullibility. Let us take up here as examples some of the rules that bear specifically on sex, music, and formal education (of course, different cults routinely decree strictures in more than in these mere three areas. Richard Stephens, in his 2015 book 'Black Sheep: The Hidden benefits of Being Bad', takes a look at a number of other taboo or frowned-upon practices and behaviours, and comes to a conclusion that *all* have at least some circumstantial or hidden benefits, leading to a view that in real life, moderation always works best.)

For free people, sex usually is an intricately complex endeavor, requiring top-level skills in self-knowledge, self-management, and more. Navigating one's own sexuality will help grow such skills, which will keep building upon progress hitherto achieved, or not, through a lifetime. Whereas the average teenager may have, say, 101 level type skills in this area, a *free* adult would have graduated to, perhaps, a 401 equivalent or so. As in everything, the obligation to follow somebody else's dictates, rather than learning and growing by oneself through a freedom which critically includes the freedom to make mistakes and learn from them, drastically impoverishes life by curtailing its associated range of choices and its breadth of experience. It quells opportunities for learning and for growth. If you're not allowed to explore something, whatever that may be, sex, violin, extreme sports, anything—you will never know whether you would have been good at it, whether you would have made the grade, or perhaps turned out to be exceptionally gifted, or just plain average or even plain hopeless. You have, for no objectively valid reason whatsoever, been denied a challenge that would have taught you something both about yourself and about the wider world you live in. You just have not been put to the test, and your life, and, because you are part of the universe, *the wider universe itself*, have not grown for it: both have been impoverished.

If some outside agency bereaves you of your freedom and ability to freely make choices, you are thereby *robbed* of the ability to face complex, grown-up choices, of the responsibilities that come with them, and of the opportunity to grow from them. The possibility of failure, of hard-knocks learning, of growth, of excellence even, have all been stolen away from you. Worse still, whenever someone is bereft of freedom, that person is also cheated of her or his ability to contribute as much as they could to stretching the envelope of the universe's manifest reality. We will revisit why, and how it leads to broader undesirable consequences.

At a more immediately practical level, many cults seem to share broadly similar stances on what their followers' proper sexual attitudes should be. Unfortunately, these striven-for attitudes often seem both directly and indirectly scientifically counterproductive and sometimes outright damaging, thus bearing out a suspicion that such stances have very little to do with godliness and all to do with dominating people, making them more malleable and pliant to the interests of the ruling classes. In terms of approved sexual attitudes, such stances may include separate categories of imposed behaviors, such as a measure of abstinence, or the curbing of opportunities to meet with members of the opposite gender.

First, abstinence. According to many a priest, intercourse should only happen for the express purpose of begetting descendants. But periods of abstinence longer than a few days have been proven to lead to starkly increased DNA damage in sperm, in turn leading to lesser fecundity and to less healthy offspring. The reservation of intercourse to only those instances when conception occurs will ineluctably lead to less healthy offspring. Similarly, limiting opportunities to find a mate from outside one's immediate circles and kin during nubile years leads to less diversity in the gene pool, with broad negative consequences.

Bearing in mind that we cannot simply transpose case studies from the animal kingdom to human societies, it is nevertheless interesting to observe what happens in the wild. Whole animal species, such as hares or guppies and some species of fish and frogs, would simply become extinct if it were not for their extreme promiscuity, a trait which in these cases amounts to a full-fledged, species-wide survival strategy. In some species, the benefits of promiscuity are both direct and individual, such as longer individual longevity rates and higher individual fecundity, *and* indirect and collective (species-wide), such as stronger offspring (51). At the other end of the spectrum, animal species such as the panda and cheetah are endangered today because, in the absence of a robust gene-shuffling strategy, the diversity of their gene pool has reduced to low and possibly non-sustainable levels. Likewise, the self-styled dictates of unwitting clerics could lead, if widely followed to the letter, to a deadly spiral of lower fecundity, poorer gene pools, and weaker descendants.

As for music, some cults view it as unwarranted, frivolous and sinful indulgence. But reality is that music is profoundly beneficial and conducive to improved performance in many areas distinct from music itself, and its use has been convincingly linked to both enhanced creative abilities and heightened cognitive performance. There is some evidence that withholding music from children may stunt brain development. Whereas the famous 'Mozart effect', a view that babies and young children who regularly listen to Mozart will grow up to become smarter adults, has not been convincingly proven, there is a large body of evidence that testifies to the positive effects of music, above all in terms of enhancing cognitive abilities as well as in therapeutic use. Amongst reams of evidence, Martin Meyer, Stefan Elmer and Lutz Jäncke have published on the positive effects that listening to music brings to the flexibility and adaptiveness of the brain's language center. Most studies, and hence the greater body of evidence, have focused on how music significantly helps children with learning difficulties or disabilities, such as certain forms of autism. Listening to, and better still learning how to play music has been proven in countless studies to significantly enhance the brain's robustness, i.e. its ability to uphold its functioning at broadly stable levels through a range of challenges (52).

Beside these studies, there is strong and varied anecdotal evidence for further benefits of music. Thomas Südhoff, a Nobel prizewinner in Physiology and Medicine, has spoken at length on how his practice of music has enabled his scientific career. Bernard Werber, a top selling French novelist, has spoken of channeling the energy he somehow taps from music into his writing, thus enhancing it. Confirming such anecdotal evidence, the psychiatrist and author Norman Doidge has written at length on how the brain can make use of music to enhance its usable energy levels. The impact of music varies depending upon the age of the listener, and indeed one of the reasons why we remain so attached to the music of our teen years is because it has helped *shape* the evolving structures of our brains in unique and permanent ways, at a developmental stage that will never happen again in our lives past our teenage years. To deprive or severely restrict music during these formative years makes the young unfolding brains less robust than they would otherwise be in the face of learning challenges. Yet this is precisely what some cults insist on doing—often together with imposing further other debilitating restrictions upon how people live their lives.

Last but not least in this brief overview, education itself. Certain cults, both historically and at the present time, restrict or even in some rare instances go to the extent of forbidding education outright. The common thread in all these approaches is an imposed-upon curbing of private life experiences—the limiting and otherwise severely restricting of the freedom of experience. But as we shall now see, mathematical analysis shows that if there is any possible Godhead-centric purpose to our lives, it would be the very opposite of restriction: the purpose would be to live as many significant or enriching experiences as possible, so as to expand the realm of actualized, as opposed to potential, reality. Such expansion would be a means to keep actualized infinity itself open-ended, ongoingly *open*. Should instant expansion cease, reality would become static—and closed. All that would then ever be left for any Godhead to experience would be well-trodden reruns. A closed reality, even if infinite, whereby all that exists and shall exist has already been visited, whereby the past and the future are equally well-known within the mind of the Godhead, where everything within reality is already a case of 'been there, done (shall do) that', could have the potential of ending divinity itself, because it would turn any Godhead who dominates space and time, and hence the past and future, into the mere curator of a kind of museum universe, where everything that is, was, or shall be in the fullness of space and time is already known within the mind of a Godhead.

To obviate the possibility of a museum universe, anything that limits, slows or prevents continued growth could properly be designated 'evil', because it would hinder the actualization and unfolding purposes of a Godhead if there is such.

It is time to look at the several layers of reality.

Part 3

The Separate Realities Within Reality

There are several layers of reality: there is actual or *manifest* reality—what was, is, or shall be out there, in terms of things that have ever existed or exist or shall exist, events that have happened or are happening or will happen at some point. Also, thoughts that have been, or which shall be, thought within some brain (intermediated by the material firing of real neurons.) It is everything that has happened, happens, or shall happen somewhere within the expanse of the universe or metaverse at some point in the fullness of time.

Then there is *potential,* unactualized reality—what could happen but has not and shall not ever.

We'll name the collection of all manifest reality anywhere and any*when* Category A, and the set of never-been actualized, never-shall-be actualized, *potential* reality, Category B. The distinction between these two categories is of such importance to the coming argument that we need to take a more in-depth look at what these two categories mean and what they may contain.

Category A reality includes for instance the fact that this morning *you were driving down the road and then you turned right.* Potential, non-actualized reality—category B—is that you could have turned left or kept going straight on (but you did turn right instead), or you might not even have been on that road at all to begin with. Category A also includes the fact that next year at some date you will definitely be driving down some particular road in a certain way. It is made up of actual '*things*' or actual events—i.e., everything that exists, has existed or shall exist at some point, whether in a material sense (e.g. a table, or some event) or abstractly (e.g. a thought, or a feeling such as what you experience when you look at a particular art painting or listen to a piece of music, and so on.) Whether concrete or abstract, some *thing* belongs to manifest reality if it has ever existed, or exists, or shall ever exist in any dimension in any of the universes that have ever existed, exist, or shall ever exist. If they exist at all, some sentient blob on Andromeda, or an 11-D baby alien in some eleven-dimensional corner of the metaverse playing with its hypercube toy (its 'hendekeract'), are all part of manifest reality. Manifest existence thus encompasses full-on material items (a house, a chocolate box, supernova SN1987A, etc.) as well as *near-material* (in the sense of not being directly material but still being immediately predicated on some materiality to exist, such as an ocean wave, a massless particle such as photon, a dream, a neuron firing, etc.), and also *fully abstract* things (such as any ancient, unwritten and now-extinct language, or modern English (53), the number π, the letter B, etc.), even though the *representation* itself of such abstract items may be material. Manifest reality includes all the *thoughts* ever held or that shall ever be held, the feelings ever felt or to be felt, the dreams ever dreamt or to be dreamed (including those that were immediately and irrevocably forgotten upon waking) by any living entity that has ever lived, lives, or will ever live anywhere and anywhen in the universe, or, if such exists, in the multiverse.

Towards the end of his life, Maurice Ravel of 'Bolero' fame kept composing and hearing in his mind what he described as 'great music'— a full-fledged Opera about Joan of Arc, for instance, and other pieces. Plagued by Pick's disease, no longer able to play the piano or even hold a pen, Ravel could not write them down or in any way utter out his mind's compositions. No one but him, in his mind's ear, ever heard nor will ever get to hear any part of these latter-day in-mind compositions. By definition of manifest reality however, these pieces absolutely do *exist*, and as such firmly belong in actual reality.

As we saw earlier, if we adopt the likely view that spacetime is granular, Category A can only be infinite if some measure of the extent of the spacetime of the universe or multiverse is infinite. This condition holds for both material and near-material infinity (for instance, an infinity of thoughts can only exist if the Bekenstein bound extends all the way to infinity, either through an infinite aggregate Bekenstein volume across the universe or metaverse (which, technically, would require an infinity of existing brains), or over infinite time. In any event, even if infinite, the cardinality of Category A is confined to a lower-end cardinality, because the infinity allowed by both infinite space if it exists and even a boundless Bekenstein volume would be of aleph-1 strength.

Category B

What category A does *not* include is anything that could have, but has never popped up, or shall never pop up onto the radar screen of reality at some point in the fullness of both space and time, including all the *possible thoughts which however have never and shall never be thought* in the fullness of time.

Category B is the realm of all possible *potential* reality: it contains all the songs that John Lennon could have written but somehow never did, all the music that Maurice Ravel could have composed either publicly or in the exclusive privacy of his own mind but never did, all the dreams that you could have possibly dreamt but never did, all the further letters that the English language could be using, over and beyond both its historical futhark alphabet and its modern stock of 26 letters, but does not, all the houses that could have been built but never were and never shall be, all the human languages which could easily have developed but did not and never will, all the novels that have never been written, all the planets that could exist but do not, and so on. Since no finite limit can possibly be set to the number of elements contained within this category, it is plainly infinite.

Interestingly, there is a line of thought that says that categories A and B are strictly equal, that within a wider multiverse all potentialities are actualized somewhere within the metaverse. They say that somewhere in the metaverse an identical copy of you turned left, that somewhere else another so far identical copy of yourself kept going straight down the street, that somewhere else a near copy of you U-turned, and so on. This argument has been made by a number of physicists, but a number of independent, compelling counter-arguments have also been made (54).

- In essence, the argument that every single item of possible, potential reality does actually happen an infinity of times within an infinite universe is based on the straightforward observation that if an object, a thought, or an event exists, then it means that the probability of its existence is not nil—it may be small, but its value is non-zero. Therefore, if the scope of the universe is infinite, everything that exists, including you, must happen or exist a number of times equal to (*the nonzero value of the probability of its existence*) multiplied by (*infinity*), which equals infinity. Presto! you do exist an infinity of times across the metaverse. This argument and line of reasoning is however false. To begin with, there is, within the particular chain of micro-events and/or elementary happenings that has led to the particular object or event or yourself coming into existence, an infinite number of choices from an infinite pool of possibilities, at many steps of the chain.

- For instance, in a continuous spacetime the construction of a particular object might depend on some angle being realized: there would be an infinity of possible exact angles within the

360 degrees of a full circle. Or, the actualization of some event might depend on a chaotic micro-event happening in a certain way at a certain time: there is an infinity of ways a random chaotic event, such as a quantum fluctuation of a given magnitude, profile and duration, can happen at a certain spot at a certain time (including in granular spacetime.)

- The final outcome of the long chain of events, embedded in the final actual object or event, is thus the end result of a chain-linked number of options chosen at many constituent steps from within an infinite pool of options, and the likelihood of this particular chain ever happening again in exactly the same way is ultimately calculable as a ratio of infinities: it is indeterminate. Furthermore, the cardinality of the an endless metaverse's cardinality is one. The cardinality of the infinite pool of options that eventually gave rise to an object, and event, or yourself may either be one (in which case the result is, as indicated above, indeterminate), or in many cases much higher than aleph one, in which case the event, object, or yourself are firmly unique across the metaverse. Other arguments supporting this view, including one narrowly predicated on time, have also been made (in *The Far Horizons of Time*, *2015.*)

The key point has to do with the respective cardinalities of these two categories A and B. A first reasonable assumption that we may make is that, even if category A is infinite, its cardinality is smaller than that of B, if only because at every step of the way, A selects out only one alternative from an infinite pool of alternative options. In both cases however, the respective cardinalities are still infinitely smaller than the aleph-infinity cardinality of Ananta. Since both Category A and Category B are of a lesser cardinality than Ananta, it ensues by definition that any Godhead is infinitely bigger than either category A or B.

There is also a straightforward proof that however immensely infinite Category B is, it cannot be of aleph-infinity cardinality, as follows. There is, beyond both categories A and B, a third category C, populated by the things that cannot possibly exist even in potentiality: it is, for instance, the songs that John Lennon would have written at the age of 400 years old, the operas written by Maurice Ravel when he was yet to be born, the battles that Julius Caesar could have won in the 21st century, and the like. Let's gather these impossible things up into some infinite set, the set Category C of not-even potential reality, of things that are impossible 'for aye'. But since Ananta encompasses all the sets in the wider metaverse, it also includes Category C, and hence is bigger than B. For every option within B, there are endlessly further alternatives within C. This means that we cannot match the constituent elements of Ananta one to one with those of Category B, and hence Category B is of a cardinality smaller than Ananta's.

By definition, Ananta infinity encompasses and includes all of the available layers of reality. In category B, and all the more so beyond B, we are firmly in the realm of non-existence, or of ghostlike, unactualized, as-yet unreal existence.

- The fact that almost all of Ananta belongs to unreality (Categories B and higher), and that Category A could well be finite, could be interpreted by some as meaning that Ananta does not exist: under that view, Ananta would not exist because the higher-cardinality parts of Ananta would firmly belong in the realms of non-existence that are categories B and higher. Beyond the circularity of the argument (Ananta does not exist because it does not exist), it does not work. Aleph-infinity cardinality was originally defined purely mathematically, on the strength of clearly existent mathematical sets. The only way the resulting contradiction (Cantor's antinomy) could be resolved was to view Ananta as exceeding the confines of mathematics alone. If we did away with the existence of aleph-infinity, the whole edifice of even ordinary mathematics would come crashing down. For instance, the hitherto infinite series of whole numbers 1,2,3,4,5 .... (which on its own ineluctably leads to aleph-infinity and to the Ananta set, and thence to the antinomy), would have to be finite. It would end at some number, and therefore there would exist some biggest whole number. But we can always add 1 to any such allegedly biggest number, and the series must remain endless, and hence Ananta must exist. We confirm its existence experimentally every day by the fact that, as far as we can tell from the little bit of math that we understand and use, math works, and we can keep our accounts in proper standing. The existence of a non-exclusively mathematical Ananta is the only trick that manages to keep the whole edifice of reality together, as Cantor was the first to recognize.

An Omega Point?

Only a comparatively extremely small part of Ananta dwells within Category A, and a Godhead, for reasons that we shall explore, might want to convert more of Itself into this category of manifest reality. This picture coincides with what Teilhard de Chardin, a French priest, had called the 'Omega Point'—a not-yet-achieved state towards which a Godhead, and, under his view, the whole of 'creation' strives. In his book *The Future of Man* (1950), he envisioned a universe steadily unfolding towards ever higher levels of complexity in both 'mindstuff' and materiality. The end point of this process would be the Omega Point, the universe's ultimate status of fully achieved and unbetterable complexity and consciousness (55).

Unrealized reality is infinitely bigger than actual, *real* reality, and it provides an infinite source capable of being tapped in order to grow reality.

- Could a multiverse, in the fullness of time, exhaust all potential actualization possibilities? Will Category B ultimately fully become submerged into A? Will all possible chocolate boxes become all actualized at some point, at least once? In other words, will the 'gilded cage' of a fully 'been-there, done-that' universe slam shut at some point? If the answer to that question is yes, the multiverse will begin to rerun at some point, and become, in effect, ever-repetitive and dull, an impossible situation for a Godhead to find Itself in.

- The answer is actually easily calculable, inferred from the different cardinalities of all possible time and space dimensions of Category A reality on the one hand, and Category B on the other hand. The cardinality of the number of points within any multidimensional spacetime is aleph-1. The number of possible functions of real numbers, or equivalently the number of potentially realizable shapes, all contained within Category B, is aleph-2. It ensues that Category B is already infinitely too big to be contained within actualized spacetime: the material multiverse cannot exhaust all potential realities.

It ensues that the number of events that can potentially be actualized within the fullness of time can only ever be a subset of the overall available number of events, unless events took zero time to happen. There thus exists an unclosable gap to stasis in any environment that includes time, such as our universe (or any other non-timeless universe.) As long as our universe (or a universe like ours) continues to exist, there is scope for category A to grow.

But how does the stuff of Category B become converted into elements of Category A reality? What are the engines of conversion and growth?

The Engines of Growth

If the endless generation of reality is going to take place thanks to universes like our own, by which mechanism does it happen? How do the formless possibilities that you are going to do this or that or maybe still something else this morning turn into the hard *fact*, a bit later on, that you have driven down the road this morning?

The Second Law of Thermodynamics stipulates a basic condition for anything to be able to 'move' and evolve in some manner from its current state or status. For any evolution or change or work to be able to take place, there must exist two separate end points or 'poles' within some associated range of attributes or properties. Examples of such poles (also called *sources*) are the negative and positive electrical poles of a battery, the north and south poles of a magnet, the 'phase lag' between two attainable electrical endpoints which enables an electrical motor to work, the two zones at different temperatures in a combustion engine that ensure that heat energy flows and therefore that some of it can be tapped off when on its way, and converted into mechanical work. Try as one may, we would not be able to extract work or movement from between two *equal* poles: everything would plain remain as is, no movement or work of any kind would occur. The second law just cannot be violated, and if the whole universe were at the exact same temperature throughout, nothing could ever happen, and the universe would be, in effect, dead (57).

More generally, life can only happen because of the productive spreads opened up by the presence of opposite poles: beauty only exists because it is set off by plainer sights, gold is valuable because there also exist other (lesser) metals, the ugliness of war occasionally opens the door to exceptional heroism and courage, poverty enables acts of generosity, and so on. In psychology, a bit of imbalance or lack of harmony can help create the tensions that motivate us and spur action. Many artists have commented on how they became creative as a means towards compensating for some perceived deficiency or shortcoming in their lives. Doris Lessing became a writer out of 'frustration', Ulay became a performance artist because of "discontent", Stephen King writes because it is, as he puts it, his way of 'keeping sane', amongst so many other such cases. Equilibrium leads to stability and stasis, whereas its lack creates dynamics, movement, work. The only reason why anybody ever gets a job is because they are able to compensate some deficiency on the part of the organization that hires them; if it were not for that shortcoming, no hire would ever occur. Life itself ultimately arises out of imbalances. If the universe were uniformly good, there would not exist any room for productive imbalances, no scope for decisions to be made, alternatives weighed, moral choices opted for: nothing much would happen there, and free will itself would degrade in such a universe.

Of course, some potentialities would be preferred over others for future actualization, namely those which, while leading to growth of reality, would skirt as much as possible the less interesting or evil bits amongst the infinite pool of available potentialities. Those deeds and actions anywhere within the multiverse which would contribute most to the expansion and growth of reality, and thereby to a measure of actualized infinity, would be *good*, actions that would slow down or even halt this process would be *evil*. Two relevant poles would be needed to generate actual reality from its potential pool. Since, in a universe featuring an infinitely good Godhead, one of the poles would embody a measure of goodness, the second pole could not possibly be all good, and the presence of some evil becomes unavoidable. Because the intensity and speed of movement also depends on the separation between poles (in much the same way that, say, a 24-volt battery is more powerful than a 9-volt), the rate of actualization of reality is boosted by a second pole further away from good.

Starkly illustrating the point, nature's ordering principle is *predation*, inasmuch as it almost exclusively drives the general evolution towards more highly complex and *better* life forms on Earth. When predatory pressure relents, biological systems tend to halt their evolution and even sometimes regress. If we view predation as, at the very least, a not very nice state of affairs—looking at the horrors that befall innocent prey at the coalface of predation, we'd be excused for thinking of it as something rather regrettable and at least mildly evil—then we discover anew a situation where the growth of something good, to wit the complexification process that leads to more capabilities, higher intelligence, and ever higher life forms, is spurred on by an opposite pole of evil. Similarly, positive human progress has leapt ahead in times of war, strife and competition, egged on by adversity and conflict. As the Orson Welles character in *The Third Man* famously (if inaccurately) put it, "in Italy, for thirty years under the Borgias, they had warfare, terror, murder and bloodshed, but they produced Michelangelo, Leonardo da Vinci and the Renaissance. In Switzerland, they had brotherly love, five hundred years of democracy and peace – and what did that produce? The cuckoo clock." In a 2014 tome, well-known historian Ian Morris explores the theme of 'War—What Is It Good For?' In a nutshell, its conclusion could be summed up by saying that as long as we are somehow not good enough, war will keep having some benefits, but that these benefits would entirely vanish the instant we become evolved enough.) Even large swathes of our culture, movies and literature feed on themes brought about by historical evil. There is no sharply delineated boundary between good and evil, and many potentialities cannot be unambiguously classified into either 'desirable' or not (for instance, you potentially could decide to rob a lazy neighbor, which could force him to become less lazy.) It is nevertheless clear that the deeper we penetrate into potential evil territory, the less desirable to a Godhead its actualization becomes.

The mere existence of our universe exists means that the ongoing actualization of potential realities is proceeding at every instant. Tomorrow morning, you will perhaps drive down that road, or choose to do something else with your time: regardless, that way, a tiny bit of Category B is being converted into A. Let us now go far into the future, a future when this universe has finally vanished. Let us assume that if they existed, all other material universes have also all dwindled to vanishing point everywhere across the metaverse. New Big Bangs, or any other kinds of universe genesis, may be brewing, but it's all 'under the surface' and for the time being Category A is but a memory in the mind of the Godhead if It exists.

In this far future, the Godhead is confined to the unreal Category B of non-reality, and all the Category A reality that has ever taken place is now closed and done with. If the Godhead wants to revisit the old realities, as It can, boring stasis has set in: these old realities have become a gilded cage of 'been there, done that, nothing new here'. Nothing *real* is ever new, nor anything new ever happens. Surprisingly, across the infinite realms of Category B, the situation is similar: the Godhead is infinitely bigger than any conceivable Category, and as such all of formless Category B is well-known for the asking. The Godhead is in danger, perhaps, of becoming relegated to areas of either static reality or of pointless unreality. The Godhead risks, perhaps, becoming... *bored*. Worse still, stasis would detract from the Godhead's full infinity, because Ananta-sized infinity is unbounded, whereas infinite stasis is bounded. By allowing stasis to go on forever, a Godhead would stand in danger of degrading Its own infinity, down from unbounded Ananta. Stephen

Hawking's famous remark that "Science doesn't prove that there is no God, only that God is not necessary", has just been turned squarely on its head: God is indeed unnecessary for the universe to exist, but maybe not the other way around. The universe, or some universe like it, seems necessary for a Godhead to not become confined into a gilded cage of been there, done that stasis. The cosmos seems to serve as a tool for a Godhead to grow Its realized infinity.

The *opposite* view to the usual believer's view of a Godhead being needed to create reality (58) seems to have now emerged. In pure science terms, our universe can full well stand on its own, without the need for any Godhead. But it turns out that a Godhead can hardly live without some universe like ours and even people like us, lest It become ... curator-like, confined into an immutably set reality and formless, unrealized realms. There we have it: because of Its sheer cardinality, a Godhead, if It exists, is inevitably mostly relegated to unactualized reality. To become ever more real, It needs manifest, material domains (of any dimensionality). Stephen Hawking's argument, often used as an argument to deny the existence of Godhead, has made a U-turn. The bottom line is that if Godhead exists, It needs reality growth to escape an ungodlike state of possible infinite stasis. Some form of activity within realms of manifest reality, of course as much as possible imbued with godlike goodness, is needed so that this growth can happen. Any activity requires the indispensable presence of a 'second pole'. This ensures that the universe cannot be uniformly and only suffused with only the Godhead's infinite traits, lest stasis sets in. A stepdown from infinity is needed in some areas so that the wider universe may grow. When applied to the quality of infinite goodness, it means that some measure of evil cannot *in principle* be fully eradicated.

In the light of this simple result, it is instructive to revisit briefly some of the dogmas which over the centuries have been invoked and promulgated by those who aver to speak on behalf of a Godhead. As it happens, some of these dogmas, because they impose restrictive lifestyles and/or curtail freedom of choice, squarely lead to a lessening of realized reality, i.e. to less Category A reality. Curbing education, forbidding or preventing experiences, dictating people how they should live their lives, promulgating blanket rules that take away their ability to make their own decisions, are all actions that lead to fewer Category B-type potentialities being converted into Category A-type reality, and therefore these actions, far from contributing to the glory of Godhead as they are often made out to be, are, by the above definition, outright *evil.*

Let's speculate a bit further here—a Godhead intent on enhancing the quality of Its realized infinity could conceivably do so by becoming a kind of *pulsating* Godhead, regularly partly branching and broadening out into materiality and availing Itself of its growth engines, then occasionally, perhaps, retracting back again to Its loftier realms where no measure of evil ever need to be countenanced. It would exist as abstract mindstuff, occasionally descending partly from higher infinity to fuel its own growth and then retreating back again—a scenario that would mesh with a number of theories that posit the existence of cyclical and/or 'bouncing universes' (59).

Should however no Godhead actually exist, the cardinality of any possible apex infinity would only take on much lower values, and indeed infinity might not even exist at all. The need for some infinity to exist, so as to help define a Godhead in the first place, no longer applies. If our visible universe constitutes the sole existing universe *and* if furthermore its spacetime is discrete, as a number of physical theories suggest, then infinity is not actualized anywhere in the material arena of objects and things (*see note* 29). If there is a multiverse and our universe is not the only extant universe, then material infinity may exist, or not, but in any event in the absence of a Godhead it could only be of a lower-ranking cardinality. Infinity, and any possible short hierarchy of physical infinities within a godless reality, would be either embedded in a multiverse, or simply in our own universe all on its lonesome (if and only if our spacetime is a continuum, in which case the apex infinity would be of aleph-1 cardinality.)

In terms of ideas, thoughts, dreams and mindstuff, the overall pool of all the thoughts of all sentient beings everywhere could not be infinite. No one can ever entirely conceptualize a whole abstract infinity (such as Euclides's infinitely long lines, or infinitely extending surfaces, or, say, a complete infinite series of numbers), if only because the whole of infinity can never be mapped onto the confines of a finite mind. The Bekenstein bound would prevent accommodating into any mind every single number from an infinite set, for example. Likewise, because of the same limitations, no non-infinite being can ever experience limitless love nor infinite goodness (which need not concern us, because we can still experience *immense* love and goodness.)

In a godless universe, there would be no bias nor pressure towards ever raising the sum total of events and experiences; the universe would be free to grow, shrink, or to eventually either halt and reverse back into some Big Crunch event or alternatively thin out and dilute. Any memory of all past history or events would eventually vanish into nothingness. We would then be living in a temporary, ultimately pointless universe or metaverse, where no higher cardinality infinities of any kind would exist.

In the absence of any Godhead, the question of why cults would have proven so resilient and sustainable over time becomes intriguing. *Something* has had to somehow keep them going. As Abraham Lincoln put it, everybody can be fooled some of the time, a few people can be fooled all of the time, but if it were not for *something* to hang their hat on, purely delusional cults based on nothing remotely tangible at all could not have fooled so many people for so long—they would simply have disappeared. What is it that kept them going? We will look at the likeliest answer below, that of the prevalence of mystical experiences, and what such experiences may mean. Before we look into what may be sustaining cults to this day, we must look at how they arose in the first place.

Historical Perspectives

Even in a godless universe there would have existed historical forces favoring the establishment of religions. Beyond providing a privileged place where individuals could go to seek explanations or solace, cults and religions have historically provided both reasons why, and venues where, the sparse or scattered communities of antiquity would come together—an essential function for people to meet and exchange and work together in common undertakings requiring the joint efforts of many—in short, for building civilization itself. Another possible contributive factor to the historical robustness of cults is the fact that, as the philosopher John G. Messerly put it, people 'do not want to know— they want to believe'. Ours are lives of permanent risk and uncertainty, and belief is a legitimate coping mechanism towards the challenges, uncertainties, and indeed the tragedies of life. To keep operating optimally we must hope that they will be useful and fulfilling, and that we will live another day. Furthermore, we all must go about the business of living our lives and making ends meet, and we do not necessarily have either the time or the luxury to delve into the deeper questions of life. A normal tendency is therefore to leave these questions to those who proclaim themselves specialists, and, often, just take their word for it.

Mono-theistic cults celebrating a single almighty Godhead appeared only relatively recently in the history of mankind. Multi-theism still endures to this day in some areas and/or cultures, and as we shall see mathematics provides an intriguing insight into what may lie beneath the many Godheads of non-monotheistic cultures. The Godheads of ancient Egypt, for instance, were many, as were those of ancient Greece and also, as far as we can ascertain, those of the many small communities that foreran the advent of the larger communities that eventually coalesced into mankind's first nation-states.

Various anthropologists and sociologists of religion, such as Ara Norenzayan, Richard Sosis, Pascal Boyer, Rodney Stark and many others, have studied how and why so many deities have appeared repeatedly in so many places, virtually in all known human communities and settlements. Their studies show that the widespread fear of all-powerful, all-seeing Godheads was the one thing that durably shifted the exclusive pursuit by individuals of their (or their clan's) own narrow self-interests towards those of the wider community. It led to community-building, and ultimately to the advent of large-scale civilization.

The science of both the historical and current role of religious feelings within communities has recently gathered pace. Archaeologists seek to understand why late Stone Age tribes began building huge temples (forerunning the construction of Europe's churches, minsters and cathedrals by thousands of years.) Psychologists conduct elaborate tests to probe the effect of faith upon attitudes, and how such attitudes may contribute or not to the commonweal of the community. What emerges is a spectrum of unexpected positive and civilization-enhancing outcomes: for instance, the building of huge temple structures (such as those at the Ness of Brodgar or Göbekli Tepe) taught people how to cooperate on large projects, for which specialization and division of labor, both foundational skills to civilization, are key. The psychologist Ara Norenzayan believes that it is first at temple building sites such as Göbekli Tepe that mankind learnt to cooperate on a significant, civilization-building scale. He further believes that the fear of a potentially wrathful Godhead is the key single element that first enabled and then spurred long-distance trade. Long trading routes were key to enabling civilization to unfold. In the absence of any technologies for oversight and control, the fear of divine retribution was the only available tool to keep dishonesty and theft at bay. (This ancient reliance on some form of religion to ensure honesty has in some cases continued unbroken to this day: the oldest Indian religion on record, Jainism, strictly demands no violence, no lying or stealing, and the several million modern-day practitioners of Jainism are renowned as unfailingly honest and reliable in their business dealings. Interestingly, Jainism is a non-theistic religion, perhaps suggesting that the implied 'all-seeing eye-in-the-sky' threat of the early theistic religions was used as a short-cut obviating the need for the lengthier personal training associated with non-theistic religions such as Jainism or Buddhism.) Further buttressing Norenzayan's views of the role of religion in forestalling dishonesty is the fact that historically, the big temples were systematically built at the nodes and junctions of the trading routes. The key role of religion thus lay in ensuring that people cooperated smoothly for the benefit of all (irrespective of any particular details of a faith, for instance whether mono- or polytheistic.)

This in turn may explain why, to this day, many governments worldwide intensely dislike atheists: they fear mavericks. All of which, of course, does not begin to prove that there is or not a Godhead, but it certainly reinforces the notion that religion has very little to do with any possible Godhead and much to do with the control and organization of society. It bolsters an idea put forward by Nancy Ellen Abrams, that instead of seemingly making up some Godhead out of whole cloth, as the builders of Göbekli Tepe and many others once did and still many seem to do today, we should rather perhaps look around the universe for what or who would qualify for the status of Godhead (we shall investigate below whether Abrams's idea is logically workable, in the light of the attributes of Godhead.)

It turns out that to sustainably establish itself and become durably successful in a worldly sense, a religion or cult must be as demanding and exacting on its followers as possible. The anthropologist Richard Sosis showed that the level and intensity of oppressive constraints imposed by a religion directly correlates positively with its worldly success. Any cult or religion that pervades all aspects of everyday life and interferes with every waking moment, in particular through imposing approved lines and patterns of thought and mandatory restrictive rules on mundane and everyday items, such as its adherents' diet, clothing, use of time, and so forth, will be much stronger than those that only demand occasional or sporadic attention (for instance on certain days only), and/or which keep largely out of the details of everyday life. Not seldom, these demands can soon totally derail out of control. As a case in point, a sect called the 'Skoptsy' in Russia during the 19th century encouraged the cutting off of sexual organs in men and breasts in females in order to foster a heightened focus on 'God'. Many of its adherents *voluntarily* and happily submitted to this rather extraordinary demand. Accordingly, the sect did not disappear on its own (but it was actively suppressed by the Soviet authorities for decades, and has virtually, if not quite, died out today.) The more pain, the more constraints, the more effort demanded of its followers, the more *unnatural*, the more successful a cult is. Enlightened, laissez-faire religions, although they might well be far closer to a true spirit of Godhood, do not succeed much in imposing a useable (in a worldly sense) level of commitment and cohesion upon their flock. The well-known subconscious psychological trait of humans that makes us subconsciously believe that if something is free, then it is of no value, seems to hold sway with a vengeance in the realm of religion. Hence, huge magnificent temples are put up, decors and costumes are made at considerable effort and expense, and then prominently used and displayed, coercive rites and practices are imposed—and simple human psychology leads to subconscious acceptance on the part of many that all this is somehow proof of the rightness of the religion or cult. Needless to say, it's nothing of the kind: however grandiose, silk costumes are just only that, i.e. purely materials things decided upon, designed and made wholly by humans. Stone temples are just that: buildings made of ordinary stones, however well-made they may be, no matter how much time and effort were expanded on building them. All the magnificence on display in any rite, all the architectural prowess and colors and Gregorian chanting and gold chalices and jewel-encrusted relics and whatever else, however lofty and uplifting and impressive and powerful and out of this world they may all seem, are all purely material and man-designed and man-made (or rather man-processed from raw materials found in nature.)

We must thus rather astonishingly conclude that mathematically untenable views of Godhood (as are still largely embedded in today's prescriptive cults and religions), have nevertheless been instrumental in building civilization from its ground zero. They staved off the selfishness and dishonesty that would otherwise have derailed any attempt at harnessing collective efforts towards the common good. This conclusion may look distasteful because we want to believe, notwithstanding overwhelming evidence to the contrary, that little good could ever stem from *bad*, that for instance honesty cannot or should not arise out of lies, delusion and falsehoods.

Should a Godhead actually exist, this conclusion seems even harder to countenance, and some try to circumvent it by positing *creationism*—an act of instant creation by a Godhead, whereby mankind's whole world and environment, complete with ready-made fossils and the like, would have been created, ready-to-go, a few thousand years ago by divine fiat. This would neatly bypass the long touch-and-go evolutionary phase of mankind when, slowly and painfully, we grew into civilized societies. This view, irrespective of any other grounds however compelling, simply cannot work: eco-systems must demonstrably evolve together, become *confluent* together, if they are to become sustainable over any period of time. The instant establishment of a perfectly balanced, fine-tuned and in principle fully sustainable eco-system has been attempted several times, in vastly different contexts—and all these attempts have consistently and quickly failed.

As a first case in point, several US and international regions have tried to artificially emulate Silicon Valley's prowess in creating so many world-beating start-ups. These attempts never went anywhere, because the correct mix of incubators, facilities, companies, financial institutions, and so on, turns out *not* to be the essential ingredient. Whenever the ecosystem is 'parachuted' down from above as is, as it were, it never works out. In order to work, the mix must have evolved organically together and found its own points of dynamic equilibrium, established its own links and relationships and feedback loops and checks and balances over time. Bottom up evolution proceeds by hits and misses, through interactive adaptation to the ways of the other players. Top down does not build robustness into necessarily complex co-dependent environments. Frederick Terman, a Stanford professor sometimes called the father of the Silicon Valley, was hired by a consortium of high tech companies to duplicate Silicon Valley's success in New Jersey, and then in Texas, and failed in both locations. Michael Porter, a Harvard Business School professor, and others also attempted to create regional innovation centers modelled after the Silicon Valley. The Silicon Valley magic could *never* be replicated.

In the animal world, similar observations have been made: whole ecosystems were replicated, modeled after fully functioning ecosystems in other areas, with the exact same mixes of flora and fauna and resources (60). In all cases, the replicates immediately foundered. Equilibrium and sustainability cannot be created from above, even by following rules and ingredients and mixes that have amply proven their worth elsewhere: they must instead establish themselves through an organic process of trial and error, and of finding their operating niches. The role of common history, of evolutive interaction with and adaptation to other players is essential.

- Further illustrating the point, the European Union was established to forestall the possibility of yet another calamitous European conflict, after Europe kept regularly going through devastating wars, beginning with the Thirty Years' War, then the Franco-Prussian War, and then two conflicts that soon blossomed into full-fledged World Wars. Let us imagine that the two World Wars never happened, and that the current institutions of the European Union were imposed onto Europe before 1914: it is extremely unlikely that such a Union of Europe imposed from without would survive. The only reason why, despite its recurrent crises and often intractable hitches and difficulties, a United Europe still endures seems to be precisely because of the various players' shared history— a partial, but essential, co-evolution. Likewise, different species of the animal world learn over time to co-adapt and co-evolve within a shared environment, and are unable to achieve sustainable co-existence in the absence of a common evolutionary history.

Regardless of the whys and wherefores of how cults began, they are still vibrantly around today, and it is extremely doubtful that they would be able to achieve this survival feat if there were not some compelling reason, at their perceived core, why they endure. As the novelist David Mitchell aptly puts it in 'The Bone Clocks', *The paranormal is persuasive: why else does religion persist?'* The existence of mystical experiences has to be the ultimate reason why all stripes of cults continue to attract and retain, no matter what, so many followers—without that connection, all cults and religions would most likely have long gone the way of the dinosaurs, likely replaced in their community-building role by more modern institutions, perhaps such as social clubs and the like.

Mystical experiences may take many forms—vivid dreams, unusual experiences, and so forth. Regrettably, many dismissive scientists are guilty of adding insult to injury when they heap undisguised scorn on people who seek explanations for what they just *know* they experienced. They know what they saw or heard or felt, and when science unduly and arrogantly belies or mocks their reality, it loses its credibility as the go-to provider of answers to life in general and to the human condition in particular. Although spiritual experiences tend by their very nature to be intensely one-off, personal and non-repeatable, and thus hardly amenable to the usual standards of scientific analysis, it should nevertheless be possible to circumscribe an envelope of possible explanations. Unfortunately, 'official' science does not bother—its reaction is most often to simply deny that such experiences exist at all, and dismiss them as aberrations, hallucinations and hoaxes—an attitude which in turn does significant harm to any respect for science as a provider of reliable answers amongst the many who have had such experiences. Coincidentally perhaps, a majority of people do not feel that scientists are trustworthy, as was recently shown by a survey conducted by the Pew Research Center on behalf of the American Association for the Advancement of Science. If science is not even able to address the reality of my experience, the thinking goes, how good can it be? Faced with the choice of trusting themselves and what they know they experienced or, say, taking the word of a distant Richard Dawkins, most people will opt to trust themselves. Worse still, by being dismissive official science opens the door to the dodgy merchants of dreams with their ulterior agendas, keen to co-opt people's spiritual lives and longings, and it lets them quietly hijack many people's view of reality. The ever-recurring presence of 'unexplainable' experiences in the life stories of so many has had, and still has, such a impact on the narratives of societies and indeed on the course of History itself that it should not be arrogantly or narrow-mindedly dismissed (61).

Of course, the unrepeatability and indeed, the unobservability of such experiences by outsiders compounds the general issue that a same piece of incontrovertible evidence can be interpreted in totally different ways, in science as well as in life. In science, an example would be Grandi's series which we saw earlier (*see note* 13), which can variously be seen as evidence that weird novel emergent phenomena occur when we shift over into infinity, or as proof that infinity does not really exist. There exist many such examples of totally different interpretations by different people of a same hard fact or of some repeatable experiment, or of the same bit of the mathematics of infinity. Yet, mathematics remains both our most objective tool of analysis bar none and, despite its occasional failures, the tool that usually commands the broadest consensus.

The Math of Mystical Experiences

The key question therefore becomes whether mystical experiences (ME) are amenable to mathematical analysis. ME's have been examined from many different angles (including by a few scientists, sometimes thereby risking in the process their reputations and careers), with interpretations ranging from uncritical acceptance through shadings of reserved judgment to denial:

- *Uncritical acceptance* can safely be dismissed outright, for all the reasons exposed prior, to wit the existence of irreconcilably contradictory versions of ME's. People who, acting on the basis of their own particular visions, founded new separate cults and movements include, in no particular order, Sergey Torop, Claude Vorilhon, Marshall Summers, Elizabeth Clare Prophet, Huynh Phu So, Wovoka, Guido von List, Akhenaten, Tamara Siuda, Franklin Albert Jones, Nakayama Miki, Zelio Fernandino de Moraes, Hong Xiuquan who founded the Taiping movement, Joseph Smith, and many others still (62). A few of these cults have led to catastrophic real-world consequences, with no discernible or long-lasting spiritual gain.

- Many of the analyses that have led to *reserved judgment* are well-argued and solid. Richard Gale for instance concludes that even if such experiences are *veridical*—meaning that they are not only truthful on the part of the people who report them, but also do reflect a genuine reality that lies outside the experiencers—they cannot be cognitive, i.e. they are inherently incapable of serving as a factual proof of such reality (broadly for the same reasons we touched upon earlier of non-repeatability and built-in privateness and inability to be communicated.) But we must add here that *even if* such experiences were not only veridical but cognitive, mathematics hints that they still would *not* prove the existence of a Godhead. Our universe may very well be a universe which features a number of higher dimensional realms (4-D and up), all of which may however be *finite*. We might be, for instance, a 3-D universe enveloping a higher 4-D universe. This could in itself explain some perceived ME's, without necessitating any divinity's agency nor existence. If higher-D universes exist, it stands to reason that they might not be empty (our own universe, a mere 3-D space, is already filled with an immense diversity of things and phenomena, all of which ultimately spring from what we called pre-reality (*see also note* 14.) Mystical experiences could conceivably somehow tap into these possible higher-D realms, and the variety of contradictory memories and renditions of the experience would then result *either* from the rendering back of broadly similar higher-D experiences onto our lower-D waking reality, *or* even conceivably from memories of different areas within such realms (much like 2-D beings visiting different areas of planet Earth would return to their 2-D environment with very different memories.)

- The *deniers* use arguments which, unfortunately for their case, are ultimately non-explanatory. A typical argument consists in attributing such experiences to misfires in the brain's circuitry variously caused by oxygen starvation, hunger, poisoning, and so on. But nowhere is it exactly explained why, how, by which precise *mechanism*, a brain affected by oxygen starvation or any other agent would become capable of producing vivid, precise experiences in the first place, nor why a sizeable number of such experiences are reported in the absence of any challenge to the brain, chemical or otherwise (63). Such explanations often do not look so much as science as the mere dressing of opinions in scientific raiments. They also sweep under the proverbial rug a number of cases where incontrovertible material consequences seemed to arise from the ME. Non-explicative 'explanations', routinely but erroneously accepted as explanatory, are not uncommon in other fields as well (64). Indeed, if ME's did not happen to exist, oxygen starvation and so on could equally be used to explain any other type of brain-intermediated experience, or even a lack of memories if it should so happen that no experiences ever arose in the circumstances which, in the real world, do occasionally give rise to mystical experiences. Whether some particular accident or trauma will give rise to an ME is unpredictable: some people in broadly similar near-death accidents remember absolutely nothing of their ordeal, whereas others wake up with tales of NDE or extraordinary ME. If instead of repeated incidences of mystical experience, brains *always* went to sleep, or, say, typically happened to revisit forgotten, long-buried past memories during a cerebral shutdown sequence, the explanation that these phenomena are generated by e.g. oxygen starvation would remain seamlessly useable, and as such, such explanations are non-explicative.

Some mystical experiences can be incontrovertibly attributed, like the John Travolta character in the 'Phenomenon' movie, to some issue in the experiencer's brain, or to drugs or hunger-induced hallucinations or any number of other such causes. Some, whatever their causes, are obviously informed by cultural references, and thereby lose all credibility as indicative of some separate and objective wider reality.

But not all can be easily dismissed. Tanya Marie Luhrmann, a professor of anthropology at Stanford, reports of such experiences, both her own and hundreds of other people's—experiences palpable enough that they sometimes leave behind material traces (as cited in Paul Davids and Gary E. Schwartz, 2016, and many such others) so that any morning-after doubts and post facto dismissals are barred. She also observes that most people tend to quickly tuck such experiences away into the corners of their minds and soon forget them, because such do not neatly fit into their prior narratives of what life is and how it works, and also because they are irrelevant to the all-important business of getting on with their daily lives. As she puts it, "Sometimes people have remarkable experiences, and then file them away as events they can't explain." Jeffrey Kripal, a professor of religion at Rice University, has come across countless reports of such events and, although he is unable to "suggest an adequate explanation for this impossible possibility", he writes: "No one has an explanation, other than, of course, the professional debunker, whose ideological denials boil down to the claim that such things never happened or, if they did, they are just anecdotes unworthy of our serious attention and careful thought. Such cowardly refusals will win nothing here but my own mocking laughter." These experiences, should we care to look for them, are everywhere. The sports scientist Mihaly Csikszentmihalyi has also reported them, as have many no-nonsense others. Bearing out a suspicion that there is something there that has not been sufficiently or properly addressed by current science, the atheist author Nancy Ellen Abrams reports of succeeding in solving a personal health matter only when she accepts, at the instigation of others, the role of an outside Godhead in her life. Once her condition is healed, she relents and rationalizes that the Godhead is in fact a part of herself, some psychological aspect of herself to be found exclusively within herself. No sooner has she settled on this explanation that the condition returns, with a vengeance—to be improved again after she re-accepts the existence of *something* spiritual bigger than herself.

In the absence of other tools, there is no going around mathematical analysis. If it provides answers, even tentative, such will be far more satisfactory than the mere sweeping of inconvenient data under the proverbial rug. And it often seems to work, including in areas where we least expect it to. For instance, there is an ample trove of field data that do not fit in within the otherwise valid conclusions that the sundry above-cited sociologists and anthropologists of religion have formulated, and these data have therefore been on the whole conveniently overlooked. For example, many European anthropologists, when embedded for some length of time in native tribes in the world's remotest places (such as Dr. Christian Rätsch in South America, and others), have reported puzzling encounters with apparent material manifestations of tribal Godheads. What gives? Are there, as Shakespeare put it, far more things in heaven and earth than are dreamt of in anyone's philosophy?

A speculative mathematical approach trying to explain these data would start from a bit of physics which we encountered earlier, namely the fact that any mind has a non-local individual wave function attached to it that evolves in time (note 43). This wave function distributes itself within geographical boundaries, with a local ultra-high probability of presence within the body, mostly in the brain that houses it. But the mind holds the power to alter the shape of the wave function—that's what free will does—and therefore can in principle create other peaks of probability of presence elsewhere if it focuses on that goal. Under the known laws of quantum physics, parts of the mind's wave function could then conceivably separate out, at least in part—*decohere*, in the technical term—and go on to evolve largely independently. So far it may seem all very theoretical, but there appears to be some sparse experimental evidence for it in the form of so-called tulpas, or thought-forms. The explorer and writer Alexandra David-Néel reports how, after hearing of Tibetan tulpas, she tried, out of curiosity, to make a tulpa out of whole cloth, by the sole means of focusing her thoughts to create one. She recounts how she felt she'd perhaps succeeded in creating one after a few weeks of intense concentration: she started seeing it around the camp. So far, it could have been a mere delusion or hallucination, but then other people, who had not been told of the experiment, started seeing it independently. As days passed, the tulpa seemed to somehow put on more substance, become more and more independent, and soon became uncontrollable and even distinctly unpleasant. Alexandra David-Néel then spent a few weeks intensely mind-focusing to try and dissolve it back into nothingness. She eventually succeeded (65).

This would suggest that at least some of the tribal gods long accepted as genuine Godheads by the respective tribes, and reportedly encountered by embedded anthropologists, could be the external manifestations of separated-out mental wave functions emanating from members of the tribe. Although such thought-forms seem relatively common and are not relegated to only a few, not necessarily wholly believable accounts of isolated travelers, they nevertheless tend to almost exclusively turn up in settings where, mostly, people believe in them: animist communities across the globe, usually away from the more materialist West. Tellingly, they are very largely absent in the West, where virtually no one cares about or believes in them. Is it a case, as the novelist Neil Gaiman writes in 'American Gods', and as the mathematics of quantum physics hints, of mind-belief itself being instrumental in first creating and then sustaining such apparitions? Whereas there no guarantee whatsoever that the foregoing is anything but pure speculation, it is, however, a more satisfactory starting point towards possible understanding than the usual peremptory dismissals or blithe accusations of mendacity by a number of mainstream scientists.

The Existential Question

Lastly we must return to the age-old question of existence. All mathematical attempts to resolve the question fail for they necessarily involve circular reasoning, because infinity demonstrably only exists in reality if it exists in reality, and it does not exist if it does not. In other words, any alleged Godhead is a non-observable, and we unavoidably wind up in a situation where perfectly sane people will think that the Godhead does not exist, and equally sane others will take the opposite view (66).

Old approaches on both sides of the existential debate have variously included attempting to demonstrate the existence of a Godhead by the *indirect* effects that could be traced back to Its presence. This however cannot be done *in principle*: any indirect godlike intervention would necessary happen at the disincarnate, pre-reality level, where all and any effects can be legitimately deemed to arise from the usual laws of physics or to spring from pure randomness. We cannot use alleged *direct* effects either. So-called miracles, such as seemingly inexplicable verified cases of healing at Lourdes, say, or reported cases of apparent divine intervention in battle, and so on, have also been occasionally drafted into this debate. The issue here is that either these cases are intensely personal and not amenable to independently verifiable audit, or that such explanations become redolent of the God of the Gaps of yore. Attempted corroboration by miracles is better left alone.

Wholly new approaches are therefore needed to try and progress the debate. As Nancy Ellen Abrams says, "If the debate—of whether a Godhead exists or not—cannot be settled, doesn't that show that it is the wrong debate?" Bingo. Abrams's solution is to look around in the universe to try and find something worthy of being called 'Godhead', and should we find it, we would define It as such and all issues would be solved.

But is it in principle doable? Definitions can easily be contrived to any pre-set degree of probability of existence: should Godhead be defined as the whole universe or metaverse, then we can say with one hundred percent certainty that it exists—Godhead as the wave function of the metaverse, as it were. If It is defined as a white-bearded, harp-thrumming old man floating about in the welkin, then we can state with one hundred percent certainty that it does not. The answer thus becomes entirely dependent on how Godhead is defined, and becomes subjective. So far, so good. As per Cantor's mathematical insight, the objective existential question can never be resolved by logic alone, a conclusion also reached by the serious philosophers who have tackled the issue from entirely different angles. Richard M. Gale for instance, after a few hundred pages of intense, broad-based analysis, concludes that "no definite conclusion can be drawn regarding the rationality of faith. Faith is subjective and outstrips our ability to reason". This matches exactly what, coming from vastly different angles, Georg Cantor and others have said, and indeed what neutral mathematical analysis itself shows. Seeking out Godhead on the basis of subjectivity is, thus, mathematically all right.

An issue, however, arises if we insist, as we must, on infinity being a property of Godhood. We are then back in a situation where all of Georg Cantor's et al. caveats hold, and existence can never be proved nor disproved. Even mathematically speaking, an *infinite* Godhead is only ever amenable to private experience, and therefore cannot in principle constitute a consensually, communally recognized Godhead, with traits and characteristics accepted and adhered to by all. If we won't settle for the lower bar of any possible least common denominator definition, then a full-fledged consensual religion celebrating a consensual infinite Godhead, whether contemplative or let alone prescriptive, is impossible. Then, whether a Godhead exists or not, there is no possible intellectually honest alternative but to rejoin Sam Harris's call for a spirituality without religion. Nancy Ellen Abrams's quest for a consensual definition hardly leaves the starting block.

- To serve community and social needs, we could seek a more modest, or degraded definition, whereby the Godhead would not be infinite. Keeping to the necessary holism that mathematics imposes (such as the property of all-knowingness, albeit in its finite rendition), we could attempt to *not* define Godhead overly precisely, both in recognition of our limitations and in a bid to meet with broad consensus, and let It be, say, an all-encompassing and holistic being. Doing so, we would still have to recognize the inescapable emergent features of such a 'being', which may in turn not meet with broad consensus. For instance, Godhead could be defined as the overarching, all-encompassing live wave function of the universe or metaverse—irrespective of whether such is finite or not. Although that wave function would definitely be seen as alive, it would not necessarily be widely perceived as a satisfactory definition, because the issue of whether it would be seen to embody a sufficient level of disruptive, ineffable divinity would not seem obvious to all. Consensus could prove, again, elusive.

Where does that leave us? Proving the *possibility* of Godhead does not even consist in trying to ascertain whether ME are for real, for even if they do exist, they prove nothing, at least as long as we keep to a definition of Godhead that requires infinity (67). Our only avenue would seem to consist in examining whether actualized infinity exists. If we could find an instance of at least one infinity of whatever cardinality in some actualized reality, then the seamless building of higher cardinalities, potentially all the way up to aleph-infinity, would become possible. As we saw earlier, we can't (68).

Be that as it may, we have little choice but to try science and logic and just follow wherever such may lead. Even here, things are less straightforward than we'd like them to be, and grains of salt will keep us open-minded: yesterday's science, once thought to be complete, turned out to be but a small subset of today's science, or rather a coarse approximation only valid within limited environments. There are enough unresolved questions in science to hint that, again, today's science will be but a small subset of tomorrow's.

Takeaways

Mathematical analysis has led to a number of unexpected insights into some of the age-old questions of faith. A first inescapable takeaway is that there exists a core contradiction at the heart of every cult and religion that just won't go away: either a Godhead is infinite and then owing to emergence there is no way It can ever be second-guessed It in any way, shape or form. Or It is not, which then by definition would reduce it to the status of some lesser Godhead (and even in this latter case, emergence would still occur.) This inescapable fact utterly destroys any directive or prescriptive religion's claim to unique truth. From the fundamental impossibility of mapping the infinite onto the finite, we are led to dismiss all the claims of those who presume to know the mind of Godhead and speak on Its behalf. The only legitimate cults therefore appear to be those which limit themselves to being *contemplative*—those which encourage their followers to try and achieve some measure of spiritual enlightenment or communion with Godhead on their own, and do not presume to tell them what It wants nor to formulate doctrines. The historical confusion and amalgamation between Godhead on the one hand and humankind's worldly prescriptive religions on the other hand has seldom been seriously called into question, reinforced as it was in society narratives by the strangest of bedfellows, each looking out for their respective interests.

Second, the correct argument that no God is needed for the Universe to exist, often used to suggest that God does not exist, has been turned on its head. The universe does not need a Godhead to exist: it's the other way around. A Godhead needs some evolving physical reality to keep unfolding—to actualize Itself ever more fully and come ever nearer to what had been termed, in the demergent language of humankind, the Omega Point. Without it, a Godhead who, as It must, dominates time and space, would be relegated to stasis—i.e., the role of curating a universe which, in the fullness of its time and space, is rigidly well-delineated—hardly a godlike situation to find Oneself in.

Third, we are led to the unexpected and counterintuitive insight that if even an *infinitely good* Godhead exists, then *some* measure of evil is unavoidable. Mathematical analysis has turned an age-old question on its head—the question of why a Godhead would tolerate evil has turned into the hard fact that a measure of evil is an inescapable price to pay for infinite reality to keep growing and pushing its boundaries and unfolding above and beyond all the places it has already reached, and ultimately for any Godhead to retain Its very quality of Godhood (as long as we accept that to retain Its status of Godhead, a Godhead cannot become constrained to a mathematically *bounded* infinity but that Its infinity must remain unbounded.)

Fourth, Godhead is a non-observable. It exists if It does, and does not if it does not, and nothing further can logically or mathematically be said about whether Godhead exists or not.

- Whether or not there is a Godhead, we are still very largely left to our own devices, to sink or swim as per our own abilities and exercise of free will, both individually and collectively (69). Underscoring the difficulties inherent to interpreting non-observables, this very circumstance is taken by many as indicative of, and by others as contradictory to, the presence of Godhead. The former theorize that we can only ever become reintegrated within divinity if we conduct ourselves, at our level, with self-started and self-sustained gravitas and worthiness. Under this view, those who are not self-starters are not ready to assume, perhaps, the larger duties of their next phase up of existence. The latter theorize that if we can't observe it, then it probably does not exist.

As noted by the sociologists of religion, cults have historically provided moral principles and guidelines, as well as fostered the advent and growth of larger communities. To some extent, they continue to provide moral guidance today, but at the steep costs of ever-recurring total meltdowns (as evidenced by the very many St. Bartholomew's Day Massacres of the world), of illogical and mendacious teachings which hinder the broad development of the critical and logical thinking so essential to tackling the challenges of a complex and competitive world, and which ultimately also detract from spirituality and from Godhead Itself if such exists.

A last remark has to do with, perhaps, requisite modesty. There are always people, on all sides of any debate, who just fiercely *know*. It is an odd experience to hear or read extremely well-argued, cogent arguments that fervently militate for one side of some debate, and then hear and read equally well-argued and well-buttressed discourses that just as eagerly and cogently argue for the other side—something we have all experienced more often than we care to remember, be it in political discussions, art reviews, or even, surprisingly, in supposedly wholly factual scientific debates. Common to all these arguments, and the ultimate reason why they clash, is that they miss the bigger picture, the wider context, the black swans that are so hard to even conceptualize. Whenever we believe too fiercely or too fanatically in anything unproven, we almost certainly suffer from a failure of imagination rather than the opposite. Here we are, with our modest 100-range IQ's and inevitably limited experience and vistas, and yet we all too often presume to be certain. Would we hold the same thoughts if our IQ's were in the 100,000 range instead, if we had one hundred separate types of color-analyzing cones in our eyes instead of the paltry three we have, if we could fly, if we had lived and learned for thousands of years instead of a few decades, if we had hundreds of trillions of neurons, and any number of other such capabilities? Would we still be fiercely convinced, or would we manage to glimpse into wider, hitherto hidden new contexts, see remote yet impacting causes we did not even realize we were blind to before? Arguments seem most often open and shut only within closed, self-contained operative environments, but neat little environments amenable to 100 IQ-type certainties are very seldom the whole tale.

EPILOGUE

At their best, unexamined beliefs have over the centuries contributed to positive outcomes: selflessness beyond people's immediate self-interests, shared efforts that ultimately laid the foundation for the rise of civilization itself, extraordinary works of music and architecture and art, and a sense of meaningful purpose for many people.

But they also seem to consistently come with an extremely dark side, not limited to particular geographical areas or religions. As yet another case in point, in 2016, a famous American minister, scion from a long and respectable lineage of preachers and ministers who advised US presidents, commented on the occasion of the 'Reason Rally', a gathering of secular Americans, that its participants faced 'eternity in Hell'—no less—for their thought-crime of not believing in any particular religion. He seemed blithely unaware of just how profoundly loathsome such a pronouncement was—that its enactment would be an act worse than even those committed by history's worst psychopaths (technically, *infinitely* worse, if it were possible, because of the 'eternity' bit.) There seems to be something in our human make-up that somehow suspends most basic judgment abilities, and curtails all critical thinking when 'religion' comes around. Specialists from many different areas of specialization have looked at why this may be so, and have come up several explanations. One attempt at understanding is put forward by the reductionist Dutch neurobiologist D.F. Swaab, who offers the intriguing explanation that the human brain has, over the millennia, evolved specific circuitry meant to ensure that young children obey their parents immediately and unconditionally in times of danger: whenever a parent says run, or hide, the child must obey forthwith to survive. Over the course of humankind's long march to the modern era, any children with a propensity to first question parents' urgent orders rather than obey them would have disproportionately turned into prey, leaving mostly those children who just obeyed, no questions asked, to live another day and eventually to beget children of their own. As the descendants of the latter, most modern humans would have inherited this circuitry. D.F. Swaab suggests that religions, by design, hijack this brain circuitry: since virtually all religious traditions tend to associate Godheads with an imagery of fathers, parents and elders, ancient child survival neural pathways trigger and are activated whenever the context of religious thought comes to the fore, leading to the brain proscribing questions.

There are many other alternative explanations. One involves the way our brain evolves during its early formative years, but not necessarily in the context of parental obedience. Human brains are especially malleable during their formative years. The brain—a blank slate at the outset—soon specializes: favored neural pathways are built, soon-to-be familiar resonance structures are laid on. Any potentially viable pathways that remain unused are not developed, and ultimately outright *cannibalized*, because biological systems are very efficient at minimizing extra effort and making use of any available or redundant resources towards their more urgent tasks at hand. You may have observed this if you ever broke a limb and had it set in a cast: even though that limb had been in continual use for years and even decades prior, the instant it became useless, the body ceased to maintain its strength and redirected its resources to other muscles and biological systems: in effect, the newly unused limb became immediately cannibalized to help support other body systems. By the time you shed the cast a few weeks later, your limb had suffered significant atrophy and was no longer functional. The same process happens in the brain. The brain features loose 'centers', i.e. areas and neural resonance networks given over to specific types of neural processing. Soon, these centers which are in habitual use set, and any abilities that have not been developed during the formative years do not have their formed centers and will be much harder to develop and lay on later on. Reinforcing that view, a whole body of studies of 'feral' children, brought up by wild animals, away from civilization and human contact, documents the difficulty of reversing the way brain centers have gelled during the formative years. Jesse Bering and many others have commented on just how the dictates and beliefs of organized religion, imposed during the formative years of childhood, may be so difficult to reverse, no matter how illogical, wrong, or harmful they may be.

Some of the other proposed explanations do not directly involve the brain. Some trace attitudes back to an inherited human need to firmly belong in some recognizable community, and to divide the world into in-groups and out-groups, with religions forming a key in-group and all else being the *other*, prone to dehumanization. Other explanations that do not involve the brain at all have also been put forward.

Be that as it may, it is certain that the way we bring up children in their formative years is hardly neutral. By instilling them with religious beliefs which may gainsay other religious beliefs inculcated to other children in other places, we are sowing the seeds of tomorrow's conflicts. And we happen to be more numerous and better-armed than ever before. As many studies demonstrate, religious families have statistically significantly more children than secular ones. This means that far more children undergo some religious training and inculcation during their formative years, than not. That would not be so dangerous if all the training everywhere essentially said the same thing, but it does not. At the latest count, there are still a few thousand religions and cults worldwide. We urgently need some acceptable, reliable way of bringing the narratives closer.

Simple science has proven largely unable to reconcile different views. As any quick survey of the literature that links spirituality with non-mathematical science shows, attempts to involve any sciences other than straightforward mathematics and its immediate derivative, mathematical physics, into the spirituality debate, routinely fail. There are a number of reasons why. First, there is little consensus as to what constitutes 'science', and even the parsing of ancient texts, in order to tease out this one or that other interpretation, is deemed to be science by some. Other sciences fall short, on a variety of grounds, sometimes because they simply never needed to delve deep enough into the nature of reality: as a case in point, until very recently, the essential feature of reality that wave functions are, was totally overlooked in the life sciences.

There are also those who insist that mathematics has nothing to do with anything spiritual or divine, that these two areas are totally separate and, by their essence, can never meet. This argument seems to be based on little more than a subconscious internalization, and rehash, of arguments propounded for various reasons by a number of cults, who then serve as judge and party in this debate. If we do not start using a modicum of math and logic, we are not only sowing the seeds of tomorrow's conflicts, but we are also teaching our children to suspend logic in the key subject matter of spirituality, still a core pursuit in many places. The suspension of judgment and logic is not neutral: it does damage to minds, and especially to young minds. We cannot expect impeccable reasoning abilities on the part of students exposed to circular reasoning and other logical errors in their catechism classes.

Can we really afford to go into the future carrying with us the baggage of ancient religions, as if, for all the evidence to the contrary, business were still 'as usual', as if so many of the old religious tropes had not only been thoroughly debunked but also proven to be extraordinarily dangerous? But how can we go about ushering some simple mathematical vetting into our spiritual narratives? Can we trust world governments to aid in this endeavour, even though many still benefit from a widespread 'fear of God' amongst the populace? The future will tell.

Further on, we might even discover hints of *who* the Godhead could possibly be. This would finally answer Nancy Ellen Abrams's call for a definition of who the Godhead is. It would have to be consistent with the mathematics that says that the Godhead is omnipresent, along with all of the flow-on consequences that stem from the properties associated with Godhood, and it could mesh with some of the cosmological models put forward by the likes of physicists Roger Penrose and many others. This Godhood would probably be someone (bearing in mind that a new emergent meaning of the word 'someone' applies here) close in form to the live and living wave function of the metaverse. But that, surely, is another tale.

End Notes

(1) Note on vocabulary used.

1.1—Because spiritual traditions have no choice but to use human language to convey what they sometimes affirm to be God's word, an important minor theme in this book will be an examination of the use of language. As we shall see, simple mathematics will help determine whether any human language can function as a possible viable tool to convey divine or godlike thoughts, and hence language is taken seriously throughout this book—including how language may have arisen, and what its power and possible limitations are. Because the word 'God' is apt to be interpreted rather differently by different people, it shall be largely shunned here, and the words *Godhead, Godhood* or *deity* used instead. Some cultures accommodate several deities, whether separately (in polytheist societies such as, say, ancient Greece), or alternatively seen as different facets of the same essence (such as the Holy Trinity in Catholicism), and hence the word deity shall be used generically, i.e. without capitalization. The term *Godhood* carries an undertone of the very *essence* of what divinity may be, common to any Godhead and to any deity, and shall be used throughout with a capitalized first letter. The word Godhead shall refer to any monotheistic God, whose relevant attributes and characteristics shall be looked at in more depth as the narrative unfolds. When referring to the Godhead or any deity, the personal pronoun describing It shall be capitalized and used in its gender-neutral rendition (*It*), as shall its corresponding possessive adjective (*Its*). In the few obvious instances where the word *infinity* is used as a synonym for Godhead, it is also capitalized.

1.2—Use of 'man': the word man (as in 'man-made' or in 'mankind') shall be mostly used to mean human (incidentally, this use is borne out by etymology, since originally, the word 'man' actually meant 'human being' rather than narrowly a 'male human'. 'Man' specifically meant 'the animal who thinks', as opposed to other animals deemed thoughtless, and unsurprisingly 'man' is akin to 'mind', and further back in time, to the word 'math' as well. The word 'human' on the other hand, etymologically meant a *male* human being: it is cognate with both the French 'homme' and the Spanish 'hombre', and ultimately with the English word 'groom' as well (in 'bridegroom', or in the now obsolete meaning of a manservant); it is also akin to the words 'humble' and 'humus', two words originally meaning 'of the Earth', seen as the abode of man, as opposed to the heavens seen as that of the gods.

1.3—The words 'cleric', 'clergy', 'priest' and similar are only ever used in the general meaning of 'priestly caste', 'men of the cloth' or other such designations, in the meaning of the various officialdoms of sundry cults, and do not refer to any particular sect, cult, or religion. The words 'cult', 'religion', 'creed', and other such words are also used generically and interchangeably, and, unless specifically stated otherwise (for instance in the reference to the Aztecs) no reference whatsoever to any particular cult or religion or denomination is ever made nor implied. Likewise, any other broadly related words such as 'church' or 'temple' are never applied here in reference to any specific Church or cult, but used generically throughout, as meaning a place of public worship, irrespective of the particular cult being conducted there.

1.4 The words 'metaverse' and 'multiverse' are used here interchangeably, and describe the possibility that our known universe is either not the only universe out there, or alternatively—and equivalently—that our universe is only a small part of a wider existing universe. In existing literature, the word 'metaverse' is sometimes used to mean other universes than our own, whereas 'multiverse' also includes our own, a distinction that we do not use here.

(2) A few individuals straddle both communities, such as George Lemaître (1894-1966), John Polkinghorne (1930-), Stanley Jaki (1924-2009), Michael Heller (1936-), Robert J. Russell (1946-), and others.

(3) The so-called 'ontological arguments' have been historical attempts to use mathematics, or at least logic, to prove the existence of a Godhead. The earliest such ontological argument was put forward by Anselm of Canterbury (1033-1109). An issue with these arguments is that they seem to all contain more or less well-hidden circular arguments, which, when stripped bare of bells and whistles, basically argue that a Godhead exists because a Godhead exists.

For instance, Norman Malcolm, a US philosopher (1911-1990), seemingly reasonably argued that:

- Either Godhead exists or it does not.

- If it does not, then it cannot come into existence. If it cannot come into existence, then its existence is impossible.

- If It exists, It cannot have come into existence, and Its existence is then necessary.

- Therefore Godhead's existence is either impossible or necessary. If impossible, then Its existence is either self-contradictory or logically absurd. Since it is neither, it follows that It exists.

This argument has been refuted at considerable length by Richard M. Gale and others. At core is a slight semantic shift of meaning in the way the word 'necessary' is used in bullet points 3 and 4. In essence, all such arguments seem to either embed a measure of circularity or of unwarranted implicit assumptions, or both. A more typical ontological argument than Malcolm's would go like this:

- Our understanding of God is a being than which no greater can be conceived.

- The idea of God exists in the mind.

- A being that exists both in the mind and in reality is greater than a being that exists only in the mind.

- If God only exists in the mind, then we can conceive of a greater being—that which exists in reality.

- We cannot imagine something that is greater than God.

- Therefore, God exists.

All this demonstration does is demonstrates that there is something, some set or collection, bigger than all else in the universe—e.g., the collection of everything that exists. It seems trivial (although it does lead to interesting complications if the set is deemed to contain itself, as we shall see in the main text.). It does not demonstrate that this set of all that is, is God.

Another rendition goes like this:

- There is existence.

- Existence is a perfection above which no perfection may be conceived.

- God is perfection and perfection in existence.

- Existence is a singular and simple reality; there is no metaphysical pluralism

- That singular reality is graded in intensity in a scale of perfection (that is, a denial of a pure monism).

- That scale must have a limit point, a point of greatest intensity and of greatest existence.

- Hence God exists.

In this rendition, the second statement constitutes the unproven statement which will ineluctably lead straight to the sought-after conclusion. There exist many such variations of ontological arguments.

(4) An attribute is anything that further qualifies or further defines something or someone: for instance, a particular car's color is an attribute of that car, a person's nationality is an attribute of that person, etc. Some attributes are permanent (a person's eye color, for instance), some may or not evolve in time (a person's political affiliation, for example), and some others constantly evolve (say, a person's age.)

(5) Yonassan Gershom , another theologian from the same cultural and theological background as Harold Kushner's, opts for an infinite Godhead and for a wholly different interpretation of why evil exists. The idea of a finite Godhead seems to lie firmly in the eye of the beholder, in personal choices and interpretations, anew underscoring the need for objective and neutral tools of analysis. The danger of unrecognized *cognitive bias* on the part of theologians and even scientists is ever-

present. As a case in point, Kushner is guilty of having subconsciously eliminated *beforehand* in his analysis possible other explanations (such as Gershom's) of why bad things sometimes happen to good people, only because these did not fit in with his *prior* worldview. *A priori* judgments without analysis are never a valid reason to eliminate possible interpretations from consideration. Such cognitive bias can be subtle yet pervasive: in his book 'A Tear at the Edge of Creation' (2013), Marcelo Gleiser argues that the search for a Theory of Everything (also called Grand Unified Theory), which underlies much research in modern theoretical physics today and is often seen as one of the 'holy grails' of modern science, stems from a subconscious cognitive bias in the minds of most physicists—perhaps triggered by a longstanding culture of monotheism in the societies and cultures whence most such physicists hail from, and that there actually are *no* objective prior grounds to believe that such a Theory of Everything should exist in the first place.

Further alternative interpretations of why evil exists have been forthcoming from other cultural angles and different spiritual traditions, and other theologians (Philip Yancey, et al.) have also given their own interpretations. As we shall see in Part 3, purely math-based analysis suggests a novel and radically different vision of why 'bad things happen': it will lead to the starkly counterintuitive insight that *if* an infinite Godhead exists, *then* a measure of evil in the universe is unavoidable. We will not further countenance here Harold Kushner's et al. concept of a finite Godhead, but on the contrary assume that it is the very quality of some infinity somewhere that can give rise to something as extraordinarily disruptive and different as Godhood, and analyze where it may lead. Indeed, infinity constitutes a disruptive break from finite reality, rather than a quality or status on a continuum, which could somehow be attained by stretching out ever 'more of the same', i.e. by augmenting whatever it is that we have available. From a more historical perspective, Wolfgang Achtner examines in his essay in 'Infinity: New Research Frontiers' the historical evolution of the concept of infinity and how it relates to Godhead, and points to a broad consensus that if It exists, Godhead must be infinite. Buttressing this view, simple straightforward combinatory analytics proves that it is not inherently possible to generate infinity from any finite base, no matter how immense this finite base may be, or be made to become. However far apart, there is no fundamental or essential cleavage between any points located on a continuous scale. Any *finite* Godhead would then necessarily exist on an uninterrupted spectrum with lowlier beings—it would be an entity continuously scalable from the status of, say, tribe chieftain. In other words, such a deity would be, for all purposes, a glorified big village mayor. On the other hand, an entity with *infinite* attributes would be something else entirely—wholly beyond the reach of, and un-constructible from, any finite environment. Its very essence would be of a wholly different nature. It would be *ineffable* as per the definition of that word, 'too great to be expressed or circumscribed by words', beyond words and indeed, as we shall see, beyond mathematics itself. Notwithstanding the few theologians who take the opposite view, the appearance of whole new levels of reality whenever infinity shows up is a necessary condition for the emergence of something as ineffable and qualitatively disruptively different from everything else as Godhead, and we will keep to the view that any alleged non-infinite Godhead would not, in fact, qualify for the status of Godhood.

Imagine for an instant that you are an immensely capable being, and you can at will visit different planets, different galaxies, different universes even, different realities and cultures and ecosystems. There are very many things and events within the continuum of our reality, or in any finite universe or metaverse, but ultimately they tend to all be variations on similar underlying themes.

You have now become capable, say, of experiencing new environments, new art, new cultures at will. You could decide to learn and soon know all about the biology of, say, exotic sea species, then you would get to know other species in-depth, the ecosystems of the Serengeti and of new types of biology from other worlds, and so forth, but soon you would realize how underlyingly similar these seemingly very different ecosystems actually are, because they all ultimately must hew to a few similar broad rules of survival and to underlying laws of biology and physics, irrespective of environment. Even if these laws are different—other universes can conceivably operate under different laws of physics—the root commonality is that all environments, however exotic, must operate under some set of laws, so that order may be brought out of chaos and the stage for life set. Across our universe, there are a gazillion ways that the extant 120-odd basic chemical elements can be combined into different molecules and from thence into elements with different properties, and all these combinations make up everything there is. Any new possible combinations are each just that: yet another variation on a few underlying themes. Whether there would exist strange universes where there would exist more base elements, or where chemical elements would be replaced by, say, arrangements in the geometry of space itself, would make little difference. In your new capacity as an immensely capable being, you would soon recognize that you could travel the world extensively, visit all of its countries, then go travel and visit other worlds, admire the high jagged sheer-wall mountains of Mars or the dense gas atmosphere of Venus, travel outward to visit other galaxies, zoom out and admire Laniakea itself (our local supercluster) from afar, and maybe even visit parallel universes operating under different laws of physics, it still remains that in the fullness of time, all of this would eventually risk becoming... *boring:* Yet another set of perhaps different, but nevertheless fundamentally similar laws of physics, another set of different lifeforms, ultimately not so different from the lifeforms you encountered before. Visiting new universes would eventually become just a repeat, with local variations, of what you would have done and experienced before. You could listen to yet another teenager's existential angst and see yet another war unfold, see yet another culture, yet another biological system, yet another specimen or rendition of this or that, yet another concert or galaxy, yet another set of physical laws, yet another Big Bang, yet another instance of transcendent love or shocking evil: you have been there, seen or done that before, and what else is new? You could read or write billions of different novels and yet, as Christopher Booker once famously noted, there are only ever seven separate basic plots. As we shall see, only the possible existence of infinity has the power to save Godhead from that ultimate *limitedness*—because it provides the disruptive level of emergence essential to the sheer inexpressible otherness embedded within the very quality of Godhood.

(6) The legitimacy of mathematics in addressing essential and/or existential questions is not a given. A few philosophers adopt the viewpoint that mathematics is wholly artificial, and hence that it cannot be used to prove any truths whatsoever—either narrowly *within* the fields of mathematics and logic, or, for the more extreme view, in any field whatever. Under such views (sometimes called the *Münchhausen trilemma*, or *Agrippa's trilemma*), the mathematical proof of any theory ultimately rests on circular reasoning, or on infinite regress, or on arbitrary and unproven assertions called axioms, and numbers and other mathematical constructs are pure figments of the mind with no counterparts in actual reality. These objections can largely be put to rest, provided a few things are heeded.

First and foremost, different parts within mathematics can rest on two very different bases: some of it, such as numbers and number theory, ultimately rests on definition: one and one is two because two is *defined* as being equal to one and one: no assumptions, no hypotheses or axioms are involved in this definition, and it ultimately does not matter whether we call two 'two' or say, deux, zwei, ni, iki, dwa, or molemmat or anything else instead: whichever label is used, the underlying reality is the same; it is not label-dependent nor subject to any assumption or interpretation. In other words, the outcome of pooling together one and one yields the same reality in any language. Oddly enough though, some people *do* take the extreme view that the concept of addition, for instance, does not reflect any external reality, but is based on axioms—in this case Peano's axioms—and is every bit axiomatic as anything in, say, geometry (see below.) We shall simply disagree with them here and affirm that, irrespective of the latter-day codification and formalization of arithmetic that was indeed put forward by the Italian mathematician Giuseppe Peano (born 1858), there *are* independent such things as numbers, addition, multiplication, subtraction or division in nature and in reality.

But numbers are not the whole tale, and there are other branches of mathematics, of which some, such as Euclidean geometry, are indeed fully built from axioms—i.e., unproven foundational assumptions (aka hypotheses), often suggested by observation of the physical world, then taken at face value and erected as a foundation upon which some whole branch of mathematics is made to rest.

Of course, no mathematical proof ultimately resting on some arbitrary or unproven axiom would actually demonstrate anything about the real world, but we will take the view that any proof only involving language labels and no assumptions, does, and stands. Very many examples can readily be found. From the instant we accept that there is an irreducible reality to the actual existence of, say, one bushel and two bushels, then we admit that there exists at least a small irreducible part to mathematics, independent of any arbitrary assumptions, the part that deals with such numbers as one and two, and that says that is a person is lent one dollar and then again one further dollar later on, then that person, neglecting any interest, factually owes two dollars.

Numbers, at the outset, look straightforward and natural enough—we do have 10 fingers, 2 eyes, and there are a countable number of trees outside the window—but soon, interesting, increasingly complex, and soon even baffling properties emerge naturally from their definitions.

For instance, prime numbers emerge naturally. Primes are numbers which are only divisible by themselves and by 1—*divisible* meaning here yielding division results which remain *whole* (such as, for instance, 7 which is divisible only by itself and by 1 to yield a whole result. Non-primes are numbers like, say, 6, which is divisible by both 3 and 2 to yield a whole result, or 21, divisible by 3 and 7.) Sprinkled throughout the infinity of whole numbers (called the *natural* numbers), there exists, alongside the majority of non-primes, an embedded infinity of primes—scattered numbers such as, say, 13, 17, ... 3181, ...*ad infinitum*. This in turn leads to all the results and theorems that flow from the sheer existence of primes. One of these flow-on problems is the so-called Riemann hypothesis, a hugely difficult, as yet never solved problem (which has to do with the way prime numbers are distributed within the wider ruck of whole numbers.) The Riemann hypothesis is one of the pending 'millennium problems'. Anyone solving it would immediately qualify for a sizeable cash prize awarded by the Clay Mathematics Institute, and win a place in History. (For a solid and very readable exposition of the problem, see Keith Devlin, 2002.)

Interestingly and most productively for our purposes, the mere existence of 1 and 2 and the other simple numbers immediately and inevitably leads to a whole *ménagerie* of other numbers: fractional numbers, irrational numbers, imaginary numbers, transcendental numbers, transfinite numbers, aleph numbers, beth numbers, and more. From a simple definition of one and two, a whole unexpectedly rich branch emerges, complex theorems arise, hugely complicated unsolved conundrums crop up—which soon include side or *meta*-questions which vastly exceed the narrow confines of pure number theory itself.

As indicated above, many other branches of math, such as geometry, are more specifically based on axioms—assumptions of the mind, less straightforward and inherently more disputable than mere vocabulary definitions. A good example would be the foundational axioms (also called postulates) of Euclidean geometry, such as 'A straight line segment can be drawn joining any two points'. This is an affirmed but unproven, and indeed unprovable, assertion, which cannot be used as such to demonstrate the essential truth of anything.

But, quite productively, the *very fact* that axioms are unproven has enormous value, as it permits to delineate an overall envelope of possibilities—i.e., where is the limit separating what is possible from what absolutely cannot be (for instance in geometry, precisely because the base axioms of Euclidean geometry are not proven but merely asserted, mathematicians have been able to fruitfully explore what happens when these axioms are held to be invalid. This analysis spawned *valid* non-Euclidean geometries, such as the Riemannian and Lobatchevsky geometries, as well as higher-dimensional geometries (which all turned out to have real-world, material uses and applicability within some environments.) Essentially, it helped to precisely scope out the domains of what is possible, since all other attempted geometries beyond those few were found to be inconsistent and impossible. Equivalently, and rather obviously, no alternate realities were ever found where one and one could be equal to three, underpinning a view that numbers do embody a reality outside of their narrow axiomatic specifications. Hadn't Carl Gauss, one of the greatest mathematicians of all time, once claimed: "Mathematics is the queen of the sciences and number theory is the queen of mathematics."?)

Areas of possibilities and impossibilities can thus be usually rather precisely scoped out, thus defining precise zones of 'impossibility' (under specific classes of axioms.) This pushes the envelope and allows for exploring possible alternate realities, including such domains as may be counterintuitive or not immediately apparent.

A further question has to do with any possible built-in, inherent limitations of mathematics itself. To wit—are there areas where mathematics itself proves that it cannot answer some mathematical question, and if so, what does it say within about the ability of mathematics to answer real-life questions?

There exist several categories of problems where the ability of mathematics to provide answers becomes questionable or even breaks down completely. For instance, in applied mathematics there exist unsolvable *computational* problems—problems that crop up in the use of math-based computer algorithms. These problems involve *applied* rather than *pure* mathematics—i.e., how math is deployed and brought to bear on specific issues. The limitations of applied math reflect limitations on the uses of math and how it can be deployed, rather than on limitations of the underlying pure math itself.

Such problems include issues of undecidability (such as the halting problem, and problems reducing to the halting problem, as follows: a simple computer program that will obviously halt, i.e. stop upon completion of its programmed task, goes like this:

01 Print 'hello'

02 End

The computer will send an order to the printer to print the word hello and then will stop.

Alternatively, a simple computer program that will obviously never end but go into an endless loop would for example go like this:

01 if 1=1 then go to 01

02 End

It is proven that it is theoretically impossible for any devisable computer algorithm to decide whether some computer programs will either halt within finite time, or alternatively continue indefinitely by going into an 'endless loop'. This is known as the halting problem. There is also no alternative way whereby we could work out a solution to this question other than by some algorithm, so the problem is truly undecidable.) Other undecidable problems include Richardson's theorem (which proves that if a mathematical expression includes certain oscillating functions, then no algorithm can ever be written that determines whether this expression can ever equal zero); the so-called Word problem (in essence, if you have one set of elements ('words') that you label or index differently by means of two separate sets of labels (or 'indices'), then there is no way to know whether two elements indexed differently are in fact the same element or not); the seemingly unprovable Collatz conjecture, and many more— all of which however have long moved into *applied math* territory, rather than pure mathematics. Whatever problems we may face in pure math tend to typically arise from our human inability to solve them, rather than from inherent limitations (such as the so-called Millennium problems, straightforward mathematical problems which are demonstrably solvable, but which, try as we may, we are unable to ever work out.)

Any deity with an infinite IQ (see *note 9*) would in principle have no difficulty in knowing the answers to this category, and indeed some of these unknowable problems can even be used to serve as an alternative definition of Godhood.

- Importantly, and underscoring the inherent lack of robustness of axioms-based mathematics, some of the statements or questions that prove to be inherently undecidable within a certain set of axioms are actually proven to be knowable under another set. For instance, the Paris-Harrington principle, (a version of a theorem called the Ramsey theorem) is undecidable within a theory of natural numbers built on the five simple axioms known as Peano's axioms, but provable in the wider system of second-order arithmetic. Samewise, Kruskal's tree theorem, which has immediate applications in computer science, is undecidable from the Peano axioms but provable in set theory. It is also worth mentioning that there can exist unexpected interactions between the numbers we use to measure external reality, and the subsequent numerical results describing that external reality. For example, the numerical value of $[\sin(\alpha)/\alpha]$, when the angle α is made to approach zero, rather surprisingly depends on which unit is used to measure the angle: if measured in radians, the limit value for $\alpha=0$ is 1 but when measured in degrees, the limit for $\alpha=0$ then takes on a value of (pi/180) or approximately 0.017453. When using numbers, we must therefore be watchful that the results of analysis do not depend on labels, words, or measuring units.

- Logical undecidability

Another category of unknowables is made up of the *logical* undecidability of certain statements, as opposed to their mere computational undecidability. These are explored in the main text because they lead straight to immediate consequences as to the nature of Godhood.

- Limits due to the Laws of Physics

A last category of unknowables are inherent physical limitations to knowledge, as can arise from quantum effects or chaotic effects. These are also looked at in the main text.

Be that as it may, mathematics has built the modern world: it is ultimately thanks to mathematics that TVs and computers and cell phones and DVDs work, that planes do not fall out of the sky, bridges hold up, and the ultimate reason why we take a myriad things for granted in modern life. In essence, we cannot have it both ways: we cannot on the one hand belie mathematics' ability to help us approach an understanding of Godhood, and/or deny the sometimes unexpected or hard-to-accept results that may stem from it in the context of an analysis of Godhood, but on the other hand happily accept its role in helping us lead modern lives (to the extent that we routinely entrust our very lives to the assumption that its laws will work, as when we ride a car or fly a plane.) We cannot keep our financial accounts in order, keep watchful track of monies owed us, and on the other hand blithely gainsay the power of mathematics to help understand Godhood, should math-based conclusions happen to clash with views we hold.

Nicholas of Cusa was the first Western theologian who in the 15th century rejected leap-of-faith, declarative approaches to theology and claimed that "mathematics is the most helpful tool in understanding God's infinity". Galileo Galilei, later on echoed by the likes of Paul Dirac, Marcus du Sautoy and others, said that "Mathematics is the language with which God has written the universe." Some others, from Pythagoras in ancient Greece to wave function realists in modern physics, have gone one step further and think that the essence of the Godhead Itself can *only* be approached as the abstract, spiritual mathematical structure which they believe ultimately constitutes the only irreducible reality.

Throughout this book, we will use straightforward mathematics, solely based on numbers, and on results that arise from a numbers-based study of infinity and infinities. Whenever there would exist doubts as to whether some mathematical result holds, or if such are based on pure axioms (as per our above definition), then we will ignore such results: we simply won't be going there. (For those readers who may be interested, the so-called Banach-Tarski paradox would constitute an example of the latter situation in the study of infinities, involving the related and contradictory two axioms of choice and of indeterminacy.)

(7) Although the traditional concept of what materialism is no longer exists. Physical stuff—the materiality made up of quarks, electrons, neutrons, protons, and so on—is now proven to be nothing but a phenomenon that emerges from the non-material *vacuum*. Since the confirmed discovery of the Higgs boson, we know that *all* of material reality emerges from the elusive quantum vacuum, i.e. vacuum at its smallest scales, aka quantum froth or sometimes also called pre-reality. For a brief exposition, see e.g. Stephen Battersby: "It's Confirmed: Matter is Merely Vacuum Fluctuations", in NewScientist, Nov. 20th, 2008.

(8) As a case in point, the cosmologist Max Tegmark argues that if we opt for the Platonic view of reality, we have then no choice but to believe, mathematically speaking, in a multiverse rather than a universe, i.e. a reality where there exists a multitude of universes beside our own (which may take on a variety of exotic traits—pocket universes, higher-dimensional universes, universes trapped beyond infinite time, and so on.) Should someone somehow simultaneously believe that *a-* there exists a Godhead and *b-* our universe is the only universe that exists, then that person would be committing a straightforward

mathematical error, a straightforward equivalent to, say, believing that one and one equals, say, four and half, or indeed a conga line. (The reverse, incidentally, does not hold: there is no immediate logical contradiction between atheism *and* the existence of a multiverse.)

(9) Technically, the concept of an infinite IQ is moot, because the IQ scale is a comparative scale within a given population (such as humankind), rather than an absolute measure. For an IQ to be properly technically infinite, its holder would have to be both part of an infinite population, and to be the one person within that population with an IQ strictly higher than everybody else's. We shall use 'IQ' here as a simple shorthand for whatever objective measure may exist.

(10) Instead of bonobos, a similar analysis could involve any other animal species. For example, ants: both human neurons and ants communicate with one another by chemical signaling. Thus, the IQ equivalent of a single individual ant can in the first instance be deemed as roughly the same as that of an individual human neuron, multiplied by some corrective factor reflecting the phenomenon of emergence, greater in the case of the human neurons than that of the ants (emergence will be discussed further on in the main text.) An average human brain has about eighty-five billion neurons; a rough and ready calculation yields an average individual ant IQ of about one millionth of a billion. We are therefore, on average, about one hundred million billion times smarter than an individual ant: a hugely big, but nonetheless *finite*, factor.

(11) Anyone put off by the spectacle of any life form, such as an ape or even an ant, wantonly hurting or victimizing its kin or any another life form, would definitely not be seen as mad nor even overly anthropomorphically oriented by standard psychiatry. Feelings of repulsion in the face of such a spectacle would fall well within the norms of standard normalcy. Rather, it is the *lack* of such feelings of revulsion which would be seen as possibly indicative of possible underlying psychological pathology.

(12) Quote from 'The Far Horizons of Time', same author. See also e.g. 'Emergence', by Steve Johnson, also 'Emergence, Complexity, and Self-Organization: Precursors and Prototypes' a collection of essays edited by Alicia Juarrero and Carl Rubino, as well as a vast number of other such resources.

(13) As noted in the quoted reference, the simple mathematical series of alternate additions and subtractions of 1, called Grandi's series, provides a straightforward illustration and proof of this phenomenon. If we add up a million matched pairs of 1 and -1, we shall quite unsurprisingly and reliably get an end value of zero: $1-1+1-1+1-1 \ldots$ (a million times....) $= 0$. If instead of a million times we repeat this addition a trillion times, or a trillion trillion times, or a googol times, or a googol times squared, we will still always get zero.

But the situation changes the instant we shift over into infinity:

Let us call G (for 'Grandi') the infinite sum of paired (+1) and (-1):

$G = 1-1+1-1+1-1+1-1 \ldots$ ad infinitum......

G is hence equal to $(1 - 1) + (1 - 1) + (1 - 1) + \ldots \ldots = 0$ (thus apparently to zero), but it is also, quite obviously, equal to: $1 + (-1 + 1) + (-1 + 1) + (-1 + 1) + \ldots = 1 + 0 = 1$ (thus equal, apparently, to 1), and we can easily also write that since $G = 1 - 1 + 1 - 1 + \ldots$, we therefore have $1 - G = 1 - \{1 - 1 + 1 - 1 + \ldots\} = 1 - 1 + 1 - 1 + \ldots = G$, therefore we obtain $(1 - G) = G$, hence $G = 0.5$.

We can keep manipulating this infinite sum G in all sorts of ways, and we'll end up with an infinity of possibly legitimate values ranging from minus infinity to plus infinity: switching to an infinity of terms in the Grandi scheme has destroyed the intuitively straightforward resulting value of zero for the sum of all the matched pairs of one's and minus one's, and replaced it with indeterminacy.

Note that some mathematicians have tried to invalidate this and other similar results by claiming that shifting around parentheses is not legitimate. There is however no mathematical justification whatsoever for this assertion, other than trying to unmathematically wiggle out of a mathematically awkward situation. Furthermore, even if accepted, this assertion would not be able to do away with the phenomenon of emergence, but would shift it sideways: since it would remain legitimate to shift around parenthesis within any *finite* arithmetic expression (as long as no mathematical mistake is made), the rule that it is not legitimate to do so in an infinite expression would amount to an *emergent* rule, appearing only within the context of infinity. Emergence would then have simply shifted from the *results* over to the *rules* of arithmetic.

(14) Depending on the respective number of dimensions in space and in time, multidimensional universes can be either stable, unstable or unpredictable (i.e., not calculable.) All universes with at least 2 dimensions of either space or time with a combined (space-like plus time-like) dimensionality count higher than 5 are inherently incalculable, and thus unknowable. All universes featuring only 1 dimension of either time or space and 4 or more dimensions of correspondingly either space or time are unstable, meaning that any such universes would quickly disappear again after having popped into existence for whatever reason. Some other categories of universes would be necessarily empty, meaning that matter as we know it would not be able to exist within such universes (matter itself being of course made up of emergent and enduring vacuum fluctuations, see note 7.)

Since more dimensions mean more latitude and freedom of movement and complexity, physical laws become immediately more complex and unpredictable in higher dimensions. For instance, depending on whether dimensionality would be odd or even, we would experience ordinary things such as sound or light very differently (assuming we'd be able to go there.) In universes with even dimensionalities of space we'd hear sound in multiple ways, much like waves on the two dimensional surface of a pond travel both outwards and inwards, so that spoken speech in such dimensionalities would become riddled with overlapping echoes and unintelligible. If matter could still somehow appear in some of these higher-D universes, the geometry of atomic arrangements would become such that all particles and elements would acquire exotic and unpredictable new properties (our ordinary metals would become gases, elementary particles would look and behave very differently to their counterparts in our own universe, etc.) This would make life impossible for lifeforms based on matter, but would not curb the ability of any non-material Godhead to be present there.

(15) Quite a similar phenomenon sometimes happens with these apparent distant cousins of alleged full-fledged mystical experiences that are ordinary, brain-generated (or brain-mediated) *dreams*. Musicians have reported dreaming symphonies of such perfect beauty that they could only ever be dreamt of in the mind, but never actualized. Hector Berlioz, for one, reports in his Memoirs of a fantastic symphony he heard in dream two nights in a row (in exactly the same rendition), and then forgot forever. Other musicians have reported hearing in their dreams what they called 'perfect pieces'—and then when they frantically strove to compose them back upon waking up, no matter how hard they tried they fell short. Yet these transcribed 'dreams', however, often turned out to become their masterpieces. Anton Bruckner dreamt the first movement of his masterpiece, symphony No. 7 in E major. He transcribed it upon awakening but despaired that he 'never got it right'. The same has happened to other musicians, including, in popular music, Keith Richards of the Rolling Stones who dreamt the music of the Rolling Stones' signature song, 'Satisfaction'. He was never happy with how it came out in the studio when he tried to recompose it, and was still striving to 'get it right' well after, unbeknownst to him, the other members of the band had already released the song—which had climbed to number one in the charts. (Keith nevertheless bitterly complained to his puzzled band mates that the song as released was not good enough.) The distortion here seems to happen because of an inherent difficulty in transposing perceived experiences that happened in *'mindscape'* reality into material reality. The phenomenon is not limited to musicians, and other creative artists have reported similar experiences. Samuel Taylor Coleridge dreamt up a whole ready-made 300-line poem, which he immediately started transcribing down upon wakening. As he was writing the 54th line, he was interrupted and called on an errand, at which point he promptly and irretrievably forgot the rest. The remaining 54 lines became his acclaimed Kubla Khan poem, and the memory-zapping interrupter has since entered the language of literature as the 'person from Porlock'. The novelist Robert Louis Stevenson dreamt most of his stories, and trained himself to better remember his dreams to keep loss of content as low as possible. Famously, the mathematical genius Srinivasa Ramanujan dreamt most of his breakthrough findings, in dreams where scrolls of complex mathematical formulas, aided by instant comprehension, somehow unfolded before his eyes. The considerable difficulty of communication and transposition between a same person's dreaming and waking states would be infinitely dwarfed by the difficulty of turning external 'godlike' realities into ordinary external reality.

(16) Because there are three dimensions of space and only one of time in our spacetime, one could be tempted to argue that 'space' should come first, as it

does. But nothing prevents the existence of spacetimes with more time-like dimensions than space-like ones (see *note 14*).

(17) We could extend this analysis to more dimensions, whereby the limitations inherent in 1-D language would become more pronounced, and language soon break down completely. We could still contrive to wring some sense out of any 3-D writing we could attempt, as follows (1-D language does not easily lend itself to added dimensions, so that any attempt to up by one dimension quickly becomes intricately contrived, and outright impossible beyond 2 extra dimensions.) Say we'd want to splice into a simple statement about two people in love with each other, an inseparable rider statement expressing the universal quality and indivisibility of non-judgmental love. The most seamless way to do that would be in 3-D. We would start from, say, Oliver's and Jennifer's love story (these names harking back to Erich Segal's eponymous novel.) First we'd write out "Oliver loves Jennifer", and then, as above, loop the text into a 2-D loop to indicate the equivalent quality of the mutual love between Oliver and Jennifer. But say that we now also want to convey the indivisibility of universal love—the participation of the whole universe in the love, as well as Oliver's and Jennifer's corresponding love for the whole of creation. Adding in the word 'all', we now write the four words Jennifer, Oliver, Loves, and All onto a 3-D sphere (this would work better in Japanese, where a same term *ai-shimasu* renders both 'love' and 'loves'. For the sake of this argument, let us use the single word 'love' to cover both 'love' and loves'.) With these four words written on a 3-D sphere, the statement can now be read in a way whereby Jennifer, Oliver, all and love are all seamlessly connected as both subjects and objects. By upping to 3-D, we have been able to build in a statement of the indivisibility of love.

(18) Any words-based ideal language would certainly contain all the words that have ever existed and quite a few million more besides, but could hardly contain an infinity of words (as we will soon see in the main text, languages not based on words can feature a simple infinity of lexicon items.) A words-based tongue therefore cannot form an infinity of finite sentences, as simple combinatory analytics demonstrates, let alone *meaningful* ones. But an infinity of possible sentences would be needed if an infinity of separate thoughts were to be conveyed, and therefore even an ideal human language would be ultimately finite and limited. In the specific case of seeking a way to say spacetime, let's say that the word for spacetime in an ideal über-tongue would be 'alpha'. Further terms would need to be defined to describe other spacetimes of *any* dimensionality—say, a spacetime of 10 dimensions in space ('space-like' dimensions) and two dimensions in time ('time-like'). In the same way as any actual human language is capable of expressing precisely *any* number within the infinite pool of possible numbers, this could still be seamlessly achieved: just like there is a precise word in English, for instance, for the number 0.0134078854377883899180189563—we could set up a vocabulary system expressing different spacetimes, whereby the above-cited spacetime would be called, say, alpha-ten-two. Yet, just like it would take a time longer than a human lifetime to merely try to enunciate or write out any sufficiently big number, the same would apply to uttering out the names of most spacetimes within an infinite (or big enough) array of spacetimes. A seamlessly capable,

better language system than a sequential, mono-channel word-based language is therefore called for. Furthermore, whenever the language would seek to express objects or concepts not directly grounded in human experience, such as features emergent within environments richer or more complex than our own, human language would not merely be impractical or unwieldy, but would inevitably fail. To conceptualize such limitations, imagine trying to describe simple color changes of a 12-D hypercube (a 'dodekeract') whereby every inner and outer inner side would not only be of a different hue, but also a combination of all the other sides' colors according to some pre-established scheme. Describing such changes can't be done by a non-mathematical language. A *matrix-like* language is called for—a multi-dimensional, multi-channel descriptor of a complex, inter-dependent reality that features all kinds of inter-related dependencies and forward, backward, sideways, and hierarchical feedback loops.

(19) For an interesting discussion of these two approaches, see Richard Nisbett (2003).

(20) A few televangelists, in particular Pat Robertson, have routinely made such claims.

(21) In e.g. op.cit. 'The Far Horizons of Time', 2015.

(22) The phrase 'an infinity of infinities', although perfectly understandable and meaningful in everyday language, does not actually make accurate sense mathematically, because, as we shall see, there exist different infinite sizes of infinity (often called *cardinalities*.)

Let's call A and B two such different-value infinities (such as different aleph numbers, which we will soon look at.) If we say that there is an infinity of infinities—a reasonable statement which is otherwise quite true – then, do we mean by it that there is:

- an A-Infinity of A-Infinities, or

- an A-infinity of B-infinities, or

- a B-infinity of B-infinities, or

- a B-infinity of A-infinities?

All those infinities are quantitatively and qualitatively different. The statement *'there is an infinity of infinities'* has therefore just become, mathematically at least, meaningless, since we cannot possibly decide whether 1), 2), 3), or 4), or some other combination of further infinities, is meant.

(We could attempt to solve the problem as follows: Instead of saying that there is an infinity of infinities, we could try to say that there is in fact an:

infinity of infinities of infinities of {.... of infinities.......} of infinities...,

with the dots standing for an infinite number of repeats of the phrase 'of infinities'. This statement would also be true, but it would also turn out to be meaningless *unless* the infinite series of infinities would somehow, owing to its infinite length, converge to a same value of infinity.) In other words, only if

- an A-infinity of A-infinities of A-infinities of {.... A-*infinities*.......} of A-infinities

is the same as

- a B-infinity of B-infinities of B-infinities of {.... B-*infinities*.......} of B-infinities

and, for that matter, the same as any mix of A- and B-infinities in the infinite series above. This is however not the case: all these infinite sums of infinities demonstrably do not mathematically have (*converge* to) the same value, *aka* the same cardinality.)

Hence, the statement 'there is an infinity of infinities' or alternatively any statement such as 'there is an infinity of infinities of infinities', containing a repetition of the word 'infinities' any number of times, are surprisingly both (fuzzily) correct and mathematically meaningless.

(23) Halves belong to a category of numbers called *rational* numbers, meaning that they can be expressed by some ratio, such as 1.5 is expressible as the ratio (3/2). In fact, we can add in *all* of the existing rational numbers into the set N without ever changing its cardinality. To graduate to the next cardinality up (aleph-one), we need to add in all the irrational numbers, i.e. the numbers that cannot be expressed exactly by some ratio, such as π or the square root of 2. Almost *all* real numbers are irrational.

(24) Starting there, we are now in a position to build ever new sets such as PS of PS-A (which we'd write out as "PS-PS-A"), and so on recurrently, so that we have a means of creating an infinity of infinite sets of ever growing cardinality. The only requirement for this to work is that the original set A be infinite.

(25) The Sanskrit word *ananta* is a compound word made up of the first element 'an-', akin to and equivalent in meaning to the English prefix un- (and the word 'no'), as well as to the Latin in- or im- (as in 'impossible'), Greek an- (as in 'anaerobic'), and of the second element 'anda', cognate with the English word 'end'. Ananta thus etymologically, as well as literally, means 'no-end'.

Any aleph number (say, aleph-x) is infinitely stronger (larger) than its immediately preceding aleph number (aleph-{x-1}) by a factor equal to itself (aleph-x). Ananta is thus Ananta-infinitely larger than any other infinity.

(26) The phrase '*strictly* smaller' means *factually* smaller, not 'either smaller or equal to'. If a quantity A is strictly smaller than a quantity B, their relationship is denoted by "A<B", whereas if A is smaller but not 'strictly smaller' than B, the relationship would be written "A≤B". Reinforcing the extraordinary quality of Cantor's antinomy, 'strictly' here actually means infinitely, because power sets are always infinitely bigger than any one of their element sets.

(27) A few people are still not fully convinced by Cantor's proof that the power set of a set is always of higher cardinality than the set itself. The resulting non-

Cantorian approaches (by the likes of e.g. Petr Vopenka, et al.), are however widely seen to suffer from severe shortcomings. Some mathematicians have even argued that a non-Cantorian set theory cannot *in principle* exist, because it would inescapably lead to more logical contradictions (Vic Dannon, et al.) A number of issues have plagued attempts to build non-Canterian alternatives. Proposed alternative theories have variously featured arbitrarily defined infinity sets, arbitrarily and indefensibly restricting the number of cardinalities to a few. Some of the weaker proposals have also exhibited a measure of circularity in their arguments.

A number of attempts have also been made to resolve the antinomy from *within* the framework of Cantorian theory, which have all led either to the same contradiction by a roundabout way, or to new intractable issues even as old ones were resolved. One noteworthy attempt was to decree that the antinomy proves that the infinite and unbounded set of all possible sets does not exist. Worth a try, because sets can be erroneously defined: outside reality has always the last say as to whether sets are properly defined. For instance, if we attempt to define two separate sets of people in a village, those who get their hair cut by the unique village barber and those who don't, then to which set does the barber belong? In the case of the antinomy, the Ananta set proves to be properly defined, and any mathematical attempt to impose otherwise also leads to mathematical contradictions and ultimately to the very meltdown that these attempts were trying to avoid.

Be that as it may, our understanding of both the mathematics and the physics of infinity is still in its infancy. Our knowledge of and approaches to other issues associated with infinities, beyond the foundational concepts of power sets and cardinality, are patchy. Beyond purely theoretical considerations pertaining to infinite sets, our understanding of infinities is still rudimentary. We know, for instance, that the infinite number of points on say, a line segment or within any higher-D hyperspace is in both cases of aleph-1 cardinality (which is itself a rather counterintuitive result, underscoring the breakdown of human intuition when dealing with the infinite.) We also know that the infinite set of all possible functions of real-numbered variables (and thence of geometrical curves) is of aleph-2 cardinality. Beyond that, we do not know of any real-world instance of aleph-3 cardinality (apart from the purely mathematical, trivial example of the power set of an aleph-2 set), let alone of, say, aleph-25 or aleph-infinity. We are still at an infinite arm's length from conversance with the infinite.

(28) There are many different ways the universe, or metaverse, could be of infinite expansion, either just so, or via an infinity of pocket universes, or through other dimensions, etc. The issue of time as a component of spacetime has been treated independently by a number of authors. Several approaches require the existence of further space dimensions beyond our usual three (of height, width and length). A number of versions of string theory require 10 space dimensions. The physicist Burkhard Heim built a comprehensive theory of reality that needs up to 12 dimensions to (possibly) work, and so-called bosonic string theory requires 26 dimensions. Time itself is mooted by some (Itzhak Bars et al.) to have more than one dimension. Some of these extra dimensions may extend

indefinitely; some others may be very small and 'curled up' (in the same way as, say, the third dimension (thickness) of a 2-D sheet of paper is present everywhere where the sheet is, but is comparatively very small.) Mathematically speaking, constructing abstract spaces or spacetimes with *any* number of dimensions is entirely straightforward—for example, the 'hypercube' equivalents to our ordinary 3-D space cube would be a so-called tesseract hypercube in 4-D, a penteract in 5-D, an enneract in 9-D, a dodekeract in 12-D, and so on. Although such constructs could conceivably assume physical reality in some spacetimes, they are very difficult to picture in one's mind. Emergence may also kick in, and the quality and configuration of further dimensions within higher dimensional spacetimes might lead to exotic and totally unforeseeable structures.

(29) Current physical theories of spacetime, including not only string theory but also a number of still-evolving theories (such as Loop Quantum Gravity, its brainchild Causal Dynamic Triangulations (CDT), Quantum Einstein Gravity, Quantum Graphity, Causal Set theory, and Noncommutative Geometry) all converge towards a picture of spacetime not being continuous but *granular* (aka discrete.)

(30) Note that, whereas every single man-made object has always been preceded by abstract thoughts about it, the reverse is not true. Unlike, perhaps, Alexandra David-Néel's tulpa (which will be covered in the main text), thinking of something does not necessarily help create it into material physicality. The thought of anything, whether an object or something purely *abstract*, such as a number unlinked to any physical object, may remain purely abstract or not, but in any event still becomes actualized into reality by the agency of that thought: all thoughts have reality, *provided they have been thought.*

(31) There are other ways whereby a number can be ushered into reality: for instance, if the number in question specifically appears, however briefly, within some calculation performed by a computer, then that number has existed for real. The number cited in the main text can be easily processed by current computer technology. Technically however, computers cannot deal with all numbers of arbitrarily length, but must truncate long decimal numbers from a certain point.

- Numbers up to a certain length can be dealt with computers based on processors featuring a high number of bits (most current computers are based on 64-bit processors, there also exist higher-end 128-bit processors and experimental 512-bit processors.) So-called concatenation algorithms also enable processors with a limited number of bits to handle numbers that would normally only be suitable for higher bit-number processors. In all cases, no processors of infinite bit count can be built, nor can any algorithm be written that would concatenate numbers to infinity within finite time, which means that the number of numbers that can ever be actualized into our reality by whatever means, brains or machines, is necessarily finite, and that there remains an infinity of numbers that can in principle never be made to appear into our reality. In other words, the immense majority of numbers, *by an infinite factor*, has never been and shall never be actualized into reality.

(32) Or, equivalently, the maximum amount of information contained in or needed to fully describe any dimensionally finite physical system, such as, say, a box, a black hole, a computer, a brain....

(33) The question of whether thoughts are generated locally within the brain or whether they may take their source remotely has been examined in various texts, such as neurologist David Eagleman's *Incognito* (2011), or, from a different angle, this author's *The Far Horizons of Time* (2015). In a nutshell, there is no consensus amongst scientists, and it might even be the wrong question, since the origin of thoughts traces back to pre-reality in both cases (see note 7).

(34) This proof is mathematically equivalent to Gödel's Incompleteness Theorem.

(35) Several texts by John Barrow, William Poundstone, Michael Brooks, Noson Yanofsky, Marcelo Gleiser, Ilya Prigogine, Dirk Hoffmann and others address at various levels the inherent limits to *human* knowledge.

(36) In technical terms, in order to perform the measurement of some observable A, the measuring system must be brought into a particular alignment, a so-called *eigenstate* of that observable, i.e. a physical state in which both the measured and the measuring entities are in sync, allowing for the measuring entity to take the full measure of the measured entity without missing any bits—without any measurement leakage. However, there is no reason why any particular eigenstate of an observable A would also happen to be an eigenstate of some other observable B. In such cases, a measuring device cannot be simultaneously brought into in alignment with both A and B. This is the situation that arises in the 'Heisenberg conjugates' situation, when two attributes cannot be measured at the same time.

The need for *some* entanglement to occur, so that measurement or observation can take place, is by the way what gives rise to the so-called 'observer effect', whereby the necessary entanglement of the *measurer* of an observable (be it a measuring machine or a live observer) with that observable, leads to a definite effect on both the observable and the observer, and thus modifies their overall respective states from what they were before the act of measurement.

(37) To know where and when to go back, the Godhead would have had to be there at the precise time juncture when some micro-event that gave rise to a subsequent chain of events took place, so we are led to a straight contradiction: the only way to be able to not present everywhere at all times would be to ... be present everywhere and at all times!

(38) For readers who may be more physics-oriented, there is a more formal way to show that an omnipresent godhead cannot be material in the usual sense, as follows: let Ψ be the wave function associated with godhead. Write that Its probability of presence is certain (probability equal to 1) in an infinite multi-dimensional metaverse (irrespective of which attributes of the metaverse

happen to be infinite), technically by integrating the conjugate product of Ψ with integration boundaries set at infinity. The only possible solution is $\Psi = 0$, meaning that the godhead cannot exist *materially*. In a *finite* metaverse the proof would not hold (the integration boundaries being then set at finite values), but then that would mean by definition of Ψ that the godhead does not possess *any* infinite attribute, thereby contradicting Its definition.

(39) To put this number in perspective, the total number of elementary particles in the known universe (the neutrons, protons and electrons that make up all of matter everywhere) is about 10 to the power 89. Illustrated another way, the known universe is approximately 95 billion light years across, or about 5×10 to the power 86 cubic centimeters. (For a derivation of the number, see William Poundstone, 1989.)

(40) Moreover, an infinite spacetime would be of aleph-1 cardinality (in terms of the number of points it would contain within itself), and therefore would not be able to accommodate the aleph-2 number of all possible geometrical curves (aka all possible functions of real numbers), which does not invalidate the discussion. Incidentally, mathematicians have recently been sounding the gap between infinity and finiteness. Some patchy but intriguing progress has been made, as follows. Proofs of mathematical statements usually either deal within mathematical objects within finite environments (e.g. Pythagoras's theorem) or infinite sets (e.g. Fermat's Last Theorem), at least one theorem (technically the Ramsey Theorem for pairs) have been shown to lead to applications within both finite and aleph-naught infinite environments.

(41) As we shall see, this statement 'incarnating fully' seems to be more a case of semantics—an artefact of language—than reality. In mathematical terms, 'incarnating fully' means 'reducing all of one's dimensionality to three', as well as 'reducing all of one's metrics to finite metrics'. To 'incarnate fully', from the standpoint of some Godhead, would then become logically equivalent to becoming a 'lesser' lifeform such as, for a human, becoming an ape or a bee: not being able to do so, or even not wanting to do so, is indispensable to remaining whole.

(42) There are a few contrary views. A few scientists believe that the firing of neurons in the brain occur over too large distances to obey quantum rules, and as such would not be able to make use of the indeterminism allowed by quantum physics. This view seems mistaken. The error stems in part from a misapprehension of the scales at which quantum effects do occur: owing to entanglement effects, as specifically reflected in Bell's inequality, quantum effects are *non local*, and, far from being restricted to micro-distances, their range is demonstrably universe-wide. It seems that there is also a misunderstanding of how time scales operate: whereas it is true that the time spans over which the firing of e.g. neurons occur are much longer than the time duration needed for a quantum event to occur, such an event can nevertheless take place at any instant within a longer span of time, and thereby lead to quantum effects. J. Al-Khalili, J. McFadden et al. have written extensively about the real-world consequences of quantum events within biological systems.

(43) A wave function is a mathematical signature, or mathematical *descriptor*, attached to everything and anything in the universe. Wave functions are attached both to individual objects (such as a single particle) or collections thereof (such as an object). They can be used to calculate the whereabouts and the properties of anything, although the calculations soon become fiendishly complicated and essentially unsolvable for anything beyond the very simplest systems (such as an electron or an individual hydrogen atom.)

Wave function realism is the view held by many physicists that wave functions embody all that can be known about reality, since all objects including whole universes are ultimately built on top of and from sub-components fully describable by wave functions, and wave functions follow a hierarchical structure. Although this view of the end-all and be-all role of wave functions as a full description of all of the reality in our universe is contested by some, it has been shown that any possible additional items that would be needed to describe reality, would have to be physical objects that operate faster than light! In other words, in *any* picture of reality, wave functions embody all of the information needed to describe anything within our universe, because any onject that would move faster than light would exist outside and beyond our universe. For a solid general exposition of related issues, see e.g. David Z Albert, Alyssa Ney (2013), et al., also 'The Far Horizons of Time' by this author, etc.

(44) The 'He' of this Chapter's title symbolizes a composite theologian (theologians have been mostly if not exclusively male throughout history), and the 'she' represents mathematics (which happens to be a feminine noun in a number of languages, such as French or German, and even sometimes in English, where several well-known mathematicians, such as Marcus du Sautoy, David Bressoud and others, have referred to mathematics as the 'Queen', rather than the King, of sciences.)

(45) Cited by Norman Doidge (2007).

(46) The British historian Edward Gibbon (1737-1794) noted in his magnum opus 'The Decline and Fall of the Roman Empire' that 'The various modes of worship which prevailed in the Roman world were all considered by the people as equally true, by the philosopher as equally false, and by the magistrate as equally useful.' Nothing has much changed. Observing the current political scene, the philosopher and author John Messerly points to 'the so-called religiosity of many contemporary politicians, whose actions belie the claim that they really believe the precepts of the religions to which they supposedly ascribe.'

(47) See e.g. '*The Unpersuadables: Adventures with the Enemies of Science*' (2014), by William Storr.

(48) See http://iheu.org/
and http://freethoughtreport.com/download-the-report/

(49) It must be noted that Buddhism (deemed by many not to be a religion in the traditional sense, perhaps because it is more 'contemplative' than prescriptive) constitutes an exception. The Dalai Lama published a long piece entitled 'Our

Faith in Science', in the November 12th 2005 edition of the New York Times, on the relationship between science and Buddhism. It included the following sentence, which set the tone for the article and reflected Buddhism's overall approach to its exploration of the nature of reality: " If science proves some belief of Buddhism wrong, then Buddhism will have to change. In my view, science and Buddhism share a search for the truth and for understanding reality. By learning from science about aspects of reality where its understanding may be more advanced, I believe that Buddhism enriches its own worldview." (Article available from the New York Times archives at: http://www.nytimes.com/2005/11/12/opinion/12dalai.html?pagewanted= all&_r=0)

(50) For an interesting discussion of the God of the Gaps, see Paul Davies, 1984.

(51) For instance, the offspring of promiscuous frogs are healthier and live longer than that of their monogamous fellows, see e.g. at *http://phys.org/news/ 2011-02-promiscuity-frog-world.html#jCp* .

Corroborating studies involving other species have been conducted by e.g. Nina Wedell et al. at the University of Exeter, Miguel Barbosa at the University of Aveiro, Anne E Magurran and Maria Dornelas at the University of St Andrews, Sean Connolly and Mizue Hisano at James Cook University, and others.

(52) The well-known journalist Sidney Stevens put together an intriguing list of the top-10 benefits of music, *see e.g. at* http://www.mnn.com/leaderboard/ stories/10-reasons-why-making-music-is-good-for-your-brain

(53) Somewhat counter-intuitively, any language such as English is a fully abstract item, since it cannot be found anywhere nor touched nor weighed nor apprehended as such in any way—all we can ever glimpse are fleeting subsets of it, or imperfect material renditions—such as dictionaries—all of which are not, as the surrealist painter René Magritte could have put it, the language itself.

(54) For an exposition of the idea that there are both infinitely many identical copies and infinitely many near-copies of yourself in an infinite multiverse, see especially 'Schrödinger's Rabbits', by Colin Bruce (2004), also David Deutsch (1998, 2012), and others. For several categories of compelling counter-arguments, see for instance Dieter Zeh (2012), Ransford, (2015), et al.

(55) The concept of an Omega Point has since been picked up by other thinkers, such as Ray Kurzweil and Frank Tipler, sometimes in different renditions and shadings.

(56) Of course, mathematics qualifies this explanation some, since if It exists, the Godhead and us are not apart but exist within a same whole.

(57) This situation is sometimes called the 'heat death' scenario, whereby there would be a uniform temperature everywhere throughout the universe, with only the tiniest of evanescent random fluctuations. This would lead to all the energy present in the universe being jelled in place, and hence no work processes could take place anywhere. In such a universe, cars could never run,

biological systems like you and me or any other biological entity could not live, and so on.

This is also the reason why, conspiracy theories about greedy energy companies notwithstanding, any all-pervading vacuum energy cannot easily be turned into a appreciable energy source: with all of space everywhere bathing in the same vacuum energy, it cannot be moved and tapped, for no cold source exists anywhere in space towards which space energy could be directed. To be able to use vacuum energy, there would have to exist a cold source somewhere in space, i.e. some spot whose vacuum energy would have to be lower than that of surrounding space. But vacuum energy is created *by space itself,* which prevents the existence of such a cold source anywhere within space. There are of course local micro-fluctuations, but nowhere near the scale that would be necessary for substantive energy harvesting (Such fluctuations do lead to local phenomena such as the so-called quantum Casimir effect, which can be used to recover tiny amounts of energy.)

Should we however try to engineer a cold spot by artificial means, we would risk unleashing no less than the end of the universe. As Joseph Lykken at the Fermi National Accelerator Laboratory notes, the end of the universe could come about by the fluke advent of some larger than usual random quantum fluctuation, just big enough to create a tiny bubble of space at a lower vacuum energy level than that of its surrounding space. This would create an instant cold source in space, out of whole cloth, as it were. Just like a football spontaneously rolls downhill, i.e. towards a location of lower potential energy, the whole universe would quickly start pouring into it: it would collapse and become *swallowed* into it, because of the bubble's lower energy state. The bubble would correspondingly rapidly swell from its original puny size and start ballooning at the speed of light, and, in the words of Dr Lykken, "sweep everything before it", and destroy our universe.

(58) Stephen Hawking's remark has often been used as a strong argument to deny the existence of a Godhead, but it is nothing of the kind. Illogically equating existence with need or usefulness towards a particular purpose, and non-existence with a lack thereof, seems astonishingly self-centered. It could stem from a melding of two ancient cognitive biases—a deep-seated utilitarian mindset born of millennia of mankind's hardscrabble existence, combining with the age-old instinctive seeking of explanations for the mysteries of nature in the agency or intervention of some godhead. Obviously, the mere fact that some supernatural agency is not needed to bring about the universe does not begin to constitute logical proof that It does not exist. Mr. Möngke is a dweller in a remote village in the north of Mongolia. He has no impact, influence, effect or agency whatsoever on the life of Mr. Smythe who happens to live in London. That does not prove that either Mr. Möngke or Smythe does not exist, and the separate facts of their existences are just two unrelated, mutually irrelevant data.

(59) As Edward Belbruno of Princeton University puts it, 'Is the universe cycling through hidden time?' A number of the credible scenarios that describe a universe headed for eventual disappearance actually describe a universe that

eventually bounces back and recreate itself, endlessly going through sequential stages of existence, disappearance, and rebirth (Roger Penrose, Paul Steinhardt, Neil Turok, Martin Bojowald, et al.)

(60) See Edward Wilson (1992), et al. Another telling illustration lies in the fact that Africa is the only continent left on Earth where there still exists a wide diversity of large animals. Humans first evolved in Africa, and co-evolved with the large animals there, leading to sustainability and the survival of both man and the large animals. Humans were a later arrival in all other continents, where no co-evolution with the native large animals had taken place, a circumstance which led to most of the big animals to be wiped out by man.

(61) Mystical experiences are independent of cultural and denominational backgrounds since they are statistically equally reported by people from all kinds of religious backgrounds, including atheists.

(62) Of course, some founders may be outright frauds intent on acquiring power or making money by exploiting the gullibility of certain members of the public. The quoted names here are however a subset of those who can reasonably be assumed to have started their movements because of visions and/or ME's they genuinely experienced and perceived as stemming from a godlike source.

(63) The same holds in the few cases when some out-of-the-ordinary experience was shared by several people, in which case the usual explanation is 'collective hallucination'. Never mind that no reliable inducing mechanism has ever been found to explain 'collective hallucination', or that the phenomenon is not repeatable, thereby failing a key requirement of objective science.

(64) To illustrate the depth of explanation sought, let us draw a parallel to explanations that are sometimes given to explain everyday things, and passed off and often perceived as sufficient, satisfactory and explanatory even when they are not. In language, the etymological origins of words provides a typical example. All too often dictionaries are quite content with explaining the source of a word by indicating that such and such word comes from, say, Old Saxon, or Latin, or some other old language. They'd state, for instance, that the word 'word' comes from Old English, or that the word 'verb' comes from the Latin. Of course, these explanations explain just well-nigh *nothing* about the word's true genesis: they only displace the question of origin by replacing the question on the origin of a word in modern English by the largely content-free explanation that the word existed in Old Saxon or some other language. But the real question of etymology is how it did arise in *any* language—the question of how it did arise in modern English cannot be truly satisfactorily answered by noting that it existed in some prior language. Old languages just did not just parachute fully formed onto the old worlds of eld any more than modern English did into our world. They bore with them a history and origins, of which the usual etymologies that pass for explanatory say exactly nothing, the ignorance as to why some word happens to exist in the English lexicon having been replaced by the equivalent ignorance of why the word happened to exist in some other prior language. Keeping with our analogy, it is ultimately content-free to say that, for instance, the word 'laugh' exists in English because it already existed

as the word 'hliehhan' in Old English, or the word 'class' exists because it existed as 'classis' in Latin—apart, perhaps, from a separate and purely non-etymological explanation as to why the word survived and/or found its way into modern English. To understand the origin of a word one has no choice but to go back all the way to the actual, *material* reason that gave rise to the stem sounds that ultimately morphed into some modern word. Sometimes, it is possible to do so and go back to whatever it was in the primitive world that gave rise to the word. More often, a word's source is lost for ever in the mists of time. A word may variously have arisen and jelled as an imitation of the rote (* defined by Webster's Dictionary as the noise of the surf crashing on the shore) of the sea, the roar of a lion, the bleat of a sheep, or the pompousness of a potentate. Let us illustrate here the level of explanation sought through one example, that of the etymology of the word 'class'.

The word 'class' ultimately arose as an imitation of the sound that newborn infants make when they cry. The sound of an infant crying is interpreted by most ears as dominated by sounds mostly heard as 'ah', 'hah' and "hlah". The prehistoric vocables for crying and thence for shouting or speaking out loud imitated the newborn's cries, ah, ha, lah, and variations thereof. Then the h sound underwent the usual changes, stayed as an h in some language renditions and turned into a 'k' in others (* For the genesis and subsequent development of language, see Marcel Locquin, Merrit Ruhlen, Guy Deutscher). The word soon evolved to describe any sound made by a voice raised above normal grunt or speech levels. The Indo-European tribes who settled in the North of Europe described a voice raised due to laughter by that same vocable harking back to an infant's raised voice, spawning the word 'laugh' itself and its cognates. They also applied the same term to the sound made by their tamed and domestic animals when they raise their voices—and the verb 'to low', describing the mooing calls of cattle, is ultimately also this same word. In the tongues of those Indo-European tribes that eventually settled in *southern* Europe, where the erstwhile prehistoric *h* sounds underwent an across-the-board shift towards *kh* and *k* sounds, the word morphed from ha and hla into a word sounding like cla and then clas, meaning a shout, also into *clamor*—the sound made by the collective loud voice of a throng (which in turn yielded '*claim*', a term still dimly bearing its original connotation of a demand made loudly). Then Rome happened, a city state dominated by a strong military culture and apparatus. Centurions were military chiefs in charge of one hundred men-at-arms. Soldiers under the orders of centurions were the minutemen of their time, rostered to respond instantly to any call for assembly and intervention. Thus, a 'shout'—a *class*, in the Roman parlance—became the description of the group of men that assembled at an instant's notice at a centurion's urgent call. Eventually, the military culture ebbed, but the word's new meaning remained: a *class*, the erstwhile shout, had by now firmly become ensconced to describe a group of people. We now have uninterruptedly traced the explanatory etymology of the word 'class' all the way back to an infant's crying in the primitive tribal makeshift shelters of yore. This explanation is real, satisfying, comprehensible, miles ahead of the poor explanation that the word comes from the Latin. And we have on the way uncovered its hitherto unsuspected cognates, words like 'laugh' and 'low', in other language branches. Although the full genesis of most of our lexicon is by now lost for ever in the mists of time, there are still many known such examples.

(65) Other phenomena seem to support the view of a mind's wave function's ability to decohere. In the early 1970s, decades after David-Néel's first publicly available report, a group of people in Toronto set out to collectively create a narrowly pre-defined thought-form, and eventually succeeded (at least in part, see 'Conjuring Up Philip', (1976), by Iris Owen and Margaret Sparrow.) Verified cases of dissociative identity disorder (*aka* split personalities) can also probably be explained by, or at the very least are accompanied by, wave function decoherence (same author, 2015)

(66) It is entirely legitimate that different people should hold different opinions on non-observables. It also applies to non-observable time: perfectly sane scientists and other people aver that time does not exist and is nothing but an abiding illusion, and equally sane others disagree. Non-observability often hints that we may be missing parts of a bigger picture, and thereby asking the wrong questions.

(67) An alternative definition could conceivably choose to designate any denizens of any higher-D realms as deities. Most monotheistic religions appear to make room for beings who are reputed to dwell in higher-D realities, firmly shut to us mere mortals: angels, devils, and so on. These beings are however not usually deemed to be Godheads as such, and neither are they by our definition, for even if they happen to exist there is no compelling reason why they would possess any infinite attributes.

(68) As we saw earlier, the *only* incontrovertible instance of actual infinity known in reality, is that of time around a blackhole. Time inside the black hole lies beyond the infinite future outside the black hole; time outside the black hole lies before the infinite past inside it. At the event horizon, the border between the black hole and its outside environment, time stops entirely: the 2-D event horizon is truly timeless. Unfortunately, time is a non-observable and no one is really sure at all whether time actually exists in any fundamental sense. There are arcane ways to try to escape this conundrum, none of which are particularly compelling

(69) There is a somewhat common view, amongst people who believe that a Godhead is taking a keen and close interest in steering the destinies of mankind, that history is largely predestined and follows a grand scheme of things, laid down by the 'shepherd of the flock' to serve its ultimate purpose. Somewhat oddly, Marxist-type materialistic thought often takes a rather similar view of the "forces of history", viewed as something akin to shifting tectonic plates, whereby individuals and statesmen and politicians are mere puppets in a wider, largely inevitable historical progression.

Observation of history however suggests it unfolds much more like chaos theory, that it is shockingly subject to the 'butterfly effect', with lone individuals operative at critical inflection points. A few spectacular cases in point:

- Two little-known individuals—Vasili Arkhipov and Stanislav Petrov—changed history totally. If someone else, with different understanding, psychologies and/or hangups, had then been in their stead, we would not be here today. For instance, if someone with the psychology of Fidel Castro had been in either Arkhipov's or Petrov's stead, overwhelming psychological evidence indicates that the outcomes would have been extremely different (for a psychological portrait of Fidel Castro, see e.g. Michael Dobbs, 2008)

- George W Bush profoundly changed world history single-handedly, in a way steered by his own personality, personal hang-ups, and educational and general knowledge background. His election was however decided on the flimsiest of bases by some 400-odd contested votes in the state of Florida (even though his opponent had a nationwide edge of some 543,000 popular votes over Bush, and there appeared to be voter registration issues in Florida, perhaps affecting thousands.) If his opponent Al Gore had won instead, it is certain that formative developments in world history since would have been very different.

- World-changing operation Barbarossa was decided by *one* individual only, against the unanimous misgivings of everyone around him. It is impossible to underestimate the effects of OB on world history.

- The observation seems to repeatedly hold at countless junctures of history, whether at a macro level (as in the case of Gavrilo Princip, who triggered World War 1) or at a determinant, micro-event level. (Some would argue that, given the circumstances and Zeitgeist at the time, the role of Gavrilo Princip in starting World War 1 could have been filled by someone else if it had not been by this particular individual, but Princip's action was ultimately only made possible by a 'perfect storm' of a long chain of unlikely happenstance events. Generally speaking, it is beyond dispute that critical watersheds in the unspooling of historical events have been determined by the particular personalities of individual bit players who just happened to be there. Another such case is that of Richard Hentsch, whose personal psychological makeup directly impacted the outcome of World War 1.)

Of course, history is also shaped by overall trends such as population growth and/or shrinkage, immigration, environmental pressures, which in turn are also conditioned by political decisions made by individuals. All-important technological progress is also conditioned by political leadership—Sam Harris muses that, had it not been for the church, the Internet could have happened in Europe sometime around the 14th century. In other words, chaotic development is alive and well. We are on our own, irrespective of whether there is a Godhead, and are—as properly befits adults—the wards of our own existence and history. Interestingly, this view of critical inflection points in History swayed by individual personalities can only convince people with a modicum of scientific bent. People choosing to interpret everything in the narrow terms of revealed faiths can still credibly insist that this chaotic progression is only apparent, that it was in fact all foreplanned and proceeding according to plan. From a pure angle of faith, this approach would not be illogical or self-contradictory: as such, it is acceptable within the narrow theoretical framework of certain faiths. What this book is primarily about is those tenets of faith that are (whether visibly or not) inconsistent and self-contradictory, not those that can stand on their own within the framework of a consistent belief system. Still, a measure of the so-called scientific approach would be called for here. The 'scientific approach' consists in testing everything against experimental reality, because of a math-based realization that even in science, theories on paper can look solid and yet be completely amiss (there are precise mathematical reasons why, e.g. see *same author*, 2015, et al.). In his 2015 book, Jerry Coyne makes the point that in every case he has observed, theoretical religious views have lost against real life experimentation whenever the two approaches suggested different outcomes.

Index of Cited Names

Further Reading

A selection of literature illuminating some aspects of the discussion, representing a range of different positions and thinking (whether or not the author agrees with them.)

Abrams, Nancy Ellen (2015).A God That Could Be Real: Spirituality, Science, and the Future of Our Planet. Beacon Press.

Baggini, Julian (2015). Freedom Regained: The Possibility of Free Will. Granta Books

Barrow, John (1999). Impossibility: The Limits of Science and the Science of Limits. Oxford University Press

Barrow, John (2012). The Book of Universes. W.W. Norton

Barrow, John (1991). Theories of Everything. Clarendon Press

Birkhead, Tim (2000). Promiscuity. Harvard University Press

Brooks, Michael (2014). At the Edge of Uncertainty. Profile Books

Bruce, Colin (2004). Schrödinger's Rabbits: The Many Worlds of Quantum. Joseph Henry Press

Carr, Bernard, *editor* (2007). Universe or Multiverse?. Cambridge University Press.

Coyne, Jerry (2015). Faith Versus Fact. Viking

Davies, Paul (1984). God and the New Physics. Simon & Schuster

Davids, Paul, & Schwartz, Gary (2016). An Atheist in Heaven. Yellow Hat Publishing.

Devlin, Keith (2002). The Millennium Problems. Basic Books

Doidge, Norman (2007). The Brain That Changes Itself . Viking

Doidge, Norman (2015). The Brain's Way of Healing: Remarkable Discoveries and Recoveries from the Frontiers of Neuroplasticity. Viking

Eagleman, David (2011). Incognito: The Secret Lives of the Brain. Pantheon Books

Ehrenreich, Barbara (2014). Living With a Wild God: A Non-Believer's Search for the Truth about Everything. Grand Central Publishing

Feierman Jay (2009).The Biology of Religious Behavior: The Evolutionary Origins of Faith and Religion. Greenwood Publishing Group

Fischer, John Martin; Kane, Robert; Pereboom, Derk; & Vargas, Manuel (2007). Four Views on Free Will. Blackwell Publishing

Gamow, George (1988) 'One Two Three . . . Infinity'. Dover Books

Gleiser, Marcelo (2010). A Tear at the Edge of Creation. Free Press (Simon & Schuster)

Gleiser, Marcelo (2014).The Island of Knowledge: The Limits of Science and the Search for Meaning. Basic Books

Gribbin, John (2009). In Search of the Multiverse. Allen Lane

Hamer, Dean (2005) The God Gene. Anchor Books

Harris, Sam (2004). The End of Faith. W.W. Norton

Harris, Sam (2014). Waking Up: A Guide to Spirituality Without Religion. Simon & Schuster

Heller, Michael, & Woodin, W. Hugh, editors (2011) Infinity: New Research Frontiers. Cambridge University Press

Johnson, Steven (2001). Emergence. Allen Lane

Kane, Robert (1998). The Significance of Free Will. Oxford University Press

Kaku, Michio (1994). Hyperspace. Oxford University Press

Kanigel, Robert (1991). The Man Who Knew Infinity: A Biography of Srinivasa Ramanujan. Charles Scribner's Sons.

Kripal, Jeffrey (2011). Authors of the Impossible: The Paranormal and the Sacred. University of Chicago Press.

Lamme, Victor (2013). De Vrije Wil Bestaat Niet. Pert Bakker (Amsterdam)

McAuliffe, Kathleen (2016). This Is Your Brain On Parasites. Houghton Mifflin Harcourt

Metzinger, Thomas (2009). The Ego Tunnel. Basic Books

Morris, Ian (2014). War: What Is It Good For? Farrar, Straus & Giroux

Nisbett, Richard (2003). The Geography of Thought. Simon & Schuster

Nørretranders, Tor (1993). Mærk Verden (The User Illusion). Gyldendal, Copenhagen.

Penrose, Roger et al. (2011) Consciousness and the Universe. Cosmology Science Publishers

Poundstone, William (1989). Labyrinths of Reason: Paradox, Puzzles, and the Frailty of Knowledge. Anchor Books

Penrose, Roger (1994). Shadows of the Mind. Oxford University Press

Prigogine, Ilya (1997). The End of Certainty: Time, Chaos and the New Laws of Nature

Ransford, H Chris (2015). The Far Horizons of Time. de Gruyter

Ransford, H Chris (2014). Godhead & Mathematics. Proceedings on the Conference between Science & Theology, Dialogo Volume 1 Issue 1, November 2014

Ransford, H Chris (2016). Immanence or Transcendence? A Mathematical View. Dialogo Journal, Volume2, Issue 2, March 2016

Rucker, Rudy (1982). Infinity and the Mind. Harvester Press Ltd

Sacks, Jonathan (2015). Not in God's Name. Hodder & Stoughton

Schmitt, Eric-Emmanuel (2015). La nuit de feu. Albin Michel, Paris.

Stapp, Henry (2011). Mindful Universe: Quantum Mechanics and the Participating Observer. Springer

Stephens, Mitchell (2014). Imagine There's No Heaven: How Atheism Helped Create the Modern World. Palgrave Macmillan Trade

Stephens, Richard (2015). Black Sheep: The Hidden Benefits of Being Bad. Hodder & Stoughton

Tegmark, Max (2014). Our Mathematical Universe. Alfred A. Knopf

Walker, Evan Harris (2000). The Physics of Consciousness: The Quantum Mind and the Meaning of Life. Perseus Publishing

Watson, Gary, *editor* (2003). Free Will (Oxford Readings in Philosophy). Oxford University Press

Williamson, Victoria (2014).You are the music: How Music Reveals What it Means to be Human. Icon Books.

Zimmer, Carl (2001). Parasite Rex. Simon & Schuster

ibidem-Verlag / *ibidem* Press
Melchiorstr. 15
70439 Stuttgart
Germany

ibidem@ibidem.eu
ibidem.eu